Fire-Induced Smoke Movement and Control

Fire-Induced Smoke Movement and Control

Editors

Chuangang Fan
Dahai Qi

MDPI • Basel • Beijing • Wuhan • Barcelona • Belgrade • Manchester • Tokyo • Cluj • Tianjin

Editors
Chuangang Fan
School of Civil Engineering
Central South University
Changsha
China

Dahai Qi
Department of Civil and
Building Engineering
University of Sherbrooke
Sherbrooke
Canada

Editorial Office
MDPI
St. Alban-Anlage 66
4052 Basel, Switzerland

This is a reprint of articles from the Special Issue published online in the open access journal *Fire* (ISSN 2571-6255) (available at: www.mdpi.com/journal/fire/special_issues/fire_induced_smoke).

For citation purposes, cite each article independently as indicated on the article page online and as indicated below:

LastName, A.A.; LastName, B.B.; LastName, C.C. Article Title. *Journal Name* **Year**, *Volume Number*, Page Range.

ISBN 978-3-0365-7317-5 (Hbk)
ISBN 978-3-0365-7316-8 (PDF)

Contents

About the Editors

Chuangang Fan

Dr. Chuangang Fan graduated from the University of Science and Technology of China and obtained his Ph.D. degree in 2015. Before that, he spent half a year as a visiting scholar at RISE Research Institutes of Sweden. In 2018, he was employed as a professor at Central South University (CSU) and, now, is the deputy dean of the department of fire protection engineering. During the past 10 years, he has been committed to the fundamental and applied research on tunnel fire dynamics and smoke control methods under extreme conditions, i.e. strong wind and heavy rainfall. He received project fundings from the National Natural Science Foundation of China (less than 20% of the funding proportion) and the Excellent Young Scientist Fund of Hunan Province. He has published 59 SCI papers as the first or corresponding author. Seven of them are ESI highly cited papers or the most cited papers in the field of tunnel fire in the last 10 years. His papers have been cited over 2500 times by many famous scholars including the international association chairman, academicians and periodical editors. He has received many rewards including the First-Class Prize of the Science and Technology Innovation Award of the China Fire Protection Association. He has given 14 oral presentations at major academic conferences, including the International Symposium on Combustion and the Asia-Oceania Symposium on Fire Science and Technology. He is the reviewer for more than 60 internationally renowned journals and conferences and has been awarded the Outstanding Reviewer award for six international journals. As a superviser, he is supervising five Ph.D. students and eight master's students, of which three of them won national scholarships (the highest level of scholarship for college students in China) and two won the President Scholarship of CSU.

Dahai Qi

Dr. Dahai Qi is an assistant professor at Université de Sherbrooke in Building Engineering. He obtained his PhD in Building Engineering at Concordia University in 2016. He was then a postdoctoral fellow at the Centre for Zero Energy Building Studies at Concordia University. He graduated from Nanjing Normal University (China) with a bachelor's degree in Building Environment and Equipment Engineering in 2006 and a Master's degree in Heating, Ventilation and Air Conditioning. In 2018, he joined the Department of Civil and Building Engineering at the Université de Sherbrooke. Dr. Qi focuses on studying resilient cooling and heating technologies to mitigate climate change impacts and building fire safety. As a Principal Investigator, he has obtained funding from the CFI-John R. Evans Leaders Fund, NSERC Discovery, FRQNT-New Research Program, etc. He co-leads Subtask C-Field Studies for the IEA-EBC Annex80 Project-Resilient Cooling of Buildings. He was and still is the guest editor of many journals, such as the *Journal of Energy and Building*, etc.

Preface to "Fire-Induced Smoke Movement and Control"

Fire-induced smoke influences the safe evacuation of occupants and firefighters' ability to extinguish a fire. About 80% of deaths in fires were caused by toxic smoke, according to statistics. Hence, how to control smoke is of great importance in order to reduce fire hazards.

In this Special Issue proposed for the journal *Fire*, which is indexed in the Science Citation Index and Scopus, we sought articles associated with fire-induced smoke movement and control in both unconfined and confined environments, including high-rise buildings, tunnels, subways, mines, atriums, street canyons, etc.

In this Special Issue, the scope was to gather original, fundamental and applied research concerning experimental, theoretical, computational and case studies that contribute towards the understanding of fire-induced smoke.

Chuangang Fan and Dahai Qi
Editors

fire

MDPI

Editorial

Preface: Special Issue on Fire-Induced Smoke Movement and Control

Chuangang Fan [1,*] and Dahai Qi [2,*]

1 School of Civil Engineering, Central South University, Changsha 410075, China
2 Department of Civil and Building Engineering, University of Sherbrooke, Sherbrooke, QC J1K 2R1, Canada
* Correspondence: chuangang.fan@csu.edu.cn (C.F.); dahai.qi@usherbrooke.ca (D.Q.)

Citation: Fan, C.; Qi, D. Preface: Special Issue on Fire-Induced Smoke Movement and Control. *Fire* **2023**, *6*, 142. https://doi.org/10.3390/fire6040142

Received: 24 March 2023
Accepted: 27 March 2023
Published: 3 April 2023

1. Introduction

Generally, fires in confined spaces have more intense burning behaviors than open-space fires due to the accumulation of heat and smoke released by fires. Driven by increasingly rapid globalization and urbanization, the immense growth of building density and complexity has led to larger fire frequency and consequences [1]. This significant increase in building fire risks have raised extensive concerns about smoke transportation and control on confined-space fires through full-scale experiments [2,3], reduced-scale experiments [4,5], numerical simulations [6,7], and statistical analysis [8]. Today, a vast number of studies on fire-induced smoke movement and control are being continuously reported. It is important to provide state-of-the-art insight and strengthen the field knowledge. This Special Issue presents a collection of 12 papers that address the fundamental and practical problems of smoke movement and control in various buildings.

2. Descriptions

This Special Issue consists of research in two directions, including smoke movement under natural ventilation and smoke control, and establishes a firm foundation for future research in the field of fire-induced smoke movement and control. Efforts regarding smoke transportation law under natural ventilation have been emphasized due to the importance of determining fire hazards. The dimension effect (curvature [5] and tunnel width [9] of underground space; deck spacing of the double-deck bridge [10]) and fire source effect (heat release [5,11] and fire source location [9]) on two-dimensional temperature distribution and smoke layer entrainment have been quantified. These scientific articles contribute fire hazard assessment methods to various building fires.

Recently, the contribution to smoke control is increasing continuously. Smoke control can be divided into passive and active control technologies. In terms of passive smoke control, a methodology for field testing of smoke control properties of fire facilities [12] and an optimized natural ventilator [13] have been provided. Compared to the passive method, active control receives more attention. Aiming at the underground space, ship engine room, and large atrium, the important design parameters of smoke control system, such as smoke vent layout [4], smoke exhaust rate [4,9,13], air supply volume [14], and longitudinal ventilation velocity [15–17], have been well-addressed. Advances in smoke exhaust strategies have the potential to guide engineering practice.

3. Future Research Direction

Although considerable progress has been made on fire-induced smoke movement and control, much work remains to be conducted. Firstly, the limitations of CFD fire modelling as the widely used method, i.e., long-lasting and costly computation, are still significant [1,18], and AI (artificial intelligence) with high-performance computing will have wider applications. Secondly, in recent years, extreme weather (e.g., heavy rainfall [19,20]) is becoming increasingly prevalent with global warming. The boundary condition describing the extreme environment will not be disregarded when studying confined space fires. Last but not least, with the complex interconnected infrastructures (e.g., underground interconnected tunnel [5,9]) and green buildings [21] rapidly emerging, the characteristics of smoke transportation in these new buildings will need to be concerned in the future.

Conflicts of Interest: The authors declare no conflict of interest.

References

1. Zhang, T.; Wang, Z.; Zeng, Y.; Wu, X.; Huang, X. Building artificial-intelligence digital fire (AID-Fire) system: A real-scale demonstration. *J. Build. Eng.* **2022**, *62*, 105363. [CrossRef]
2. Long, Z.; Liu, C.; Yang, Y.; Qiu, P.; Tian, X.; Zhong, M. Full-scale experimental study on fire-induced smoke movement and control in an underground double-island subway station. *Tunn. Undergr. Space Technol.* **2020**, *103*, 103508. [CrossRef]
3. He, J.; Huang, X.; Ning, X.; Zhou, T.; Wang, J.; Yuen, R. Stairwell smoke transport in a full-scale high-rise building: Influence of opening location. *Fire Saf. J.* **2020**, *117*, 103151. [CrossRef]
4. Tao, L.; Zeng, Y. Effect of different smoke vent layouts on smoke and temperature distribution in single-side multi-point exhaust tunnel fires: A case study. *Fire* **2022**, *5*, 28. [CrossRef]
5. Tao, H.; Xu, Z.; Zhou, D. Investigation of the temperature beneath curved tunnel ceilings induced by fires with natural ventilation. *Fire* **2022**, *5*, 90. [CrossRef]
6. Jafari, S.; Farhanieh, B.; Afshin, H. Numerical investigation of critical velocity in curved tunnels: Parametric study and establishment of new model. *Tunn. Undergr. Space Technol.* **2023**, *135*, 105021. [CrossRef]
7. Liu, Q.; Xiao, J.; Cai, B.; Guo, X.; Wang, H.; Chen, J.; Zhang, M.; Qiu, H.; Zheng, C.; Zhou, Y. Numerical simulation on the effect of fire shutter descending height on smoke extraction efficiency in a large atrium. *Fire* **2022**, *5*, 101. [CrossRef]
8. Zhang, Y.; Huang, X. A review of tunnel fire evacuation strategies and state-of-the-art research in China. *Fire Technol.* **2022**. [CrossRef]
9. Liu, Q.; Xu, Z.; Xu, W.; Tagne, S.M.S.; Tao, H.; Zhao, J.; Ying, H. Study of the heat exhaust coefficient of lateral smoke exhaust in tunnel fires: The effect of tunnel width and transverse position of the fire source. *Fire* **2022**, *5*, 167. [CrossRef]
10. An, W.; Shi, L.; Wang, H.; Zhang, T. Study on the effect of bridge deck spacing on characteristics of smoke temperature field in a bridge fire. *Fire* **2022**, *5*, 114. [CrossRef]
11. Lin, W.; Liu, Q.; Zhang, M.; Cai, B.; Wang, H.; Chen, J.; Zhou, Y. Numerical simulation on smoke temperature distribution in a large indoor pedestrian street fire. *Fire* **2023**, *6*, 115. [CrossRef]
12. Hung, H.; Lin, C.; Chuang, Y.; Luan, C. Application development of smoke leakage test apparatus for door sets in the field. *Fire* **2022**, *5*, 12. [CrossRef]
13. Li, M.; Qiang, Y.; Wang, X.; Shi, W.; Zhou, Y.; Yi, L. Effect of wind speed on the natural ventilation and smoke exhaust performance of an optimized unpowered ventilator. *Fire* **2022**, *5*, 18. [CrossRef]
14. Wu, X.; Zhang, Y.; Jia, J.; Chen, X.; Yao, W.; Lu, S. Experimental and theoretical analysis of the smoke layer height in the engine room under the forced air condition. *Fire* **2023**, *6*, 16. [CrossRef]
15. Xu, Z.; Zhen, Y.; Xie, B.; Tagne, S.M.S.; Zhao, J.; Ying, H. Study on the effect of blockage ratio on maximum smoke temperature rise in the underground interconnected tunnel. *Fire* **2023**, *6*, 50. [CrossRef]
16. Na, W.; Chen, C. A study on the evacuation spacing of undersea tunnels in different ventilation velocity conditions. *Fire* **2022**, *5*, 48. [CrossRef]
17. You, W.; Kong, J. Feasibility analysis of cross passage ventilation and smoke control in extra-long submarine tunnel. *Fire* **2022**, *5*, 102. [CrossRef]
18. Wu, X.; Zhang, X.; Huang, X.; Xiao, F.; Usmani, A. A real-time forecast of tunnel fire based on numerical database and artificial intelligence. *Build. Simul.* **2022**, *15*, 511–524. [CrossRef]
19. Fan, C.; Luan, D.; Bu, R.; Sheng, Z.; Wang, F.; Huang, X. Can heavy rainfall affect the burning and smoke spreading characteristics of fire in tunnels? *Int. J. Heat Mass Transf.* **2023**, *207*, 123972. [CrossRef]
20. Luan, D.; Bu, R.; Sheng, Z.; Fan, C.; Huang, X. Experimental study on the impact of asymmetric heavy rainfall on the smoke spread and stratification dynamics in tunnel fires. *Tunn. Undergr. Space Technol.* **2023**, *134*, 104992. [CrossRef]
21. Zhang, X.; Aram, M.; Qi, D.; Wang, L. Numerical simulations of smoke spread during solar roof fires. *Build. Simul.* **2022**, *15*, 561–570. [CrossRef]

fire

MDPI

Article

Study on the Effect of Bridge Deck Spacing on Characteristics of Smoke Temperature Field in a Bridge Fire

Weiguang An [1,2,3], Lei Shi [1,2], Hailei Wang [4] and Taike Zhang [4,*]

1 Jiangsu Key Laboratory of Fire Safety in Urban Underground Space, China University of Mining and Technology, Xuzhou 221116, China
2 Key Laboratory of Gas and Fire Control for Coal Mines, China University of Mining and Technology, Ministry of Education, Xuzhou 221116, China
3 State Key Laboratory of Coal Resources and Safe Mining, China University of Mining and Technology, No. 1 University Road, Xuzhou 221116, China
4 Guang Dong Bay Area Traffic Construction Investment Co., Ltd., Guangzhou 510699, China
* Correspondence: ztke@foxmail.com

Abstract: The numerical simulation method is used to simulate the distribution characteristics of the smoke temperature field of a double-deck bridge smoke temperature field during tanker fire under natural ventilation. The influence of the distance between double decks on the truss and ceiling temperature field change in the double-deck bridge is investigated. The results show that the range of high-temperature area gradually decreases with the increase in bridge deck spacing. The maximum excess temperature function of the tunnel ceiling is also applicable to the bridge, but the coefficient is smaller than that of the tunnel experimental formula. An equation is proposed to predict the maximum excess temperature of the truss under different bridge deck spacings. As the bridge deck spacing increases, the maximum excess temperature decreases. The excess temperature of the truss increases along the truss, and the maximum excess temperature appears at the top of the truss. Based on the energy equation, an equation for the excess temperature of the truss is established. As the vertical height increases, the excess temperature of the truss above the fire source exponentially increases. The research results will contribute to the fire hazard evaluation and safety design of bridges.

Keywords: double-deck bridge; tanker fire; FDS numerical simulation; maximum excess temperature; temperature distribution

Citation: An, W.; Shi, L.; Wang, H.; Zhang, T. Study on the Effect of Bridge Deck Spacing on Characteristics of Smoke Temperature Field in a Bridge Fire. *Fire* **2022**, *5*, 114. https://doi.org/10.3390/fire5040114

Academic Editors: Chuangang Fan and Dahai Qi

Received: 26 July 2022
Accepted: 8 August 2022
Published: 12 August 2022

1. Introduction

Bridges are an important part of modern city construction. Once a fire hazard occurs, it will cause great losses. On 31 August 2004, a car burst into flames on a double-deck road bridge in Imphal in northeast India, causing an explosion that paralyzed the surrounding road system and caused extensive damage. In a fire event, the mechanical properties of the bridge decrease rapidly as the temperature increases. Therefore, it is vital to study the temperature variation characteristic of bridges under fire hazard.

The influence of fire hazards on bridges has been previously discussed by scholars. Garlock et al. [1] presented a detailed review of actual fire incidents, case studies related to fire hazards, and post-fire assessment and repair strategies for bridges. Their study pointed out that the number of damaged bridges caused by fire is nearly 3 times more than that caused by earthquakes. Peris-Sayol [2] analyzed information related to 154 cases of bridge fire, proposed classifying the damage levels suffered by a bridge during fire, and established the main factors involved in bridge fire damage. Mendes [3] and Fernando [4] et al. conducted a numerical simulation of a ship fire after the cable pylon of the Vasco da Gama bridge was hit by ships and obtained the temperature field and fire resistance time of the main beam section of the cable-stayed bridge under such a fire scenario. Bennetts et al. [5] conducted a simulation analysis on three fire scenarios for cable-stayed bridges without fire prevention design of the main tower, providing the actual time when the main tower was subjected to different ratios of ultimate load with or without fire prevention. Smith et al. [6] conducted a risk and vulnerability assessment (RVA) during the design and construction of four cable-stayed bridges in the United States by considering main tower protection, bridge deck stability, cable loss, fire prevention, and other factors, and put forward different mitigation decisions. Although the current design guidelines lack specific suggestions on how to deal with different security-related risks, designers need to ensure that the risk assessment level is enough. Ataei et al. [7] adopted the nonlinear finite element modeling and analysis method to study the influence of hypothetical fire and temperature gradient propagation along a cable and studied the influence of fire intensity and fire duration on cable strength loss by using the finite element method.

Accordingly, although many scholars have studied bridge fire, most research focuses on the influence of high-temperature fire on mechanical properties. For steel bridges, however, the critical buckling stress of the bridge is reduced greatly due to the rapid increase in temperature. This study focuses on a steel bridge in Guangzhou, China, i.e., the Shiziyang bridge, which is in the design stage. Fire Dynamics Simulator (FDS) is used to simulate the smoke temperature field of this double-deck bridge during fire to study the temperature variation of the ceiling and truss. On the one hand, the variation rule of the smoke temperature field with the change of bridge deck spacing is obtained in this paper. It is helpful for designers to choose the best bridge deck spacing for the Shiziyang bridge to achieve a balance between fire safety and economy. The temperature field of different components and different positions of the bridge is also obtained, which provides guidance for the Shiziyang bridge to adopt a zoning fire resistance scheme, i.e., different fire resistance strategies for different zones. On the other hand, there is no relevant code concerning bridge fire protection in China. There is also no basis for fire detection, fire resistance design, and safe evacuation, which are closely related to the smoke temperature field. Therefore, this study can provide basic theories and data for the establishment of bridge fire codes.

2. Methods and Models of Numerical Simulation

2.1. Methods and Models

2.1.1. Methods and Models of FDS

FDS is a powerful fire simulator software developed by the National Institute of Standards and Technology of America. This software is very flexible and widely used in the field of fire, and it can predict a variety of substances such as smoke and carbon monoxide. The accuracy of FDS has been verified by a large number of experiments [8].

FDS is used for governing equations, which are solved via the numerical method [9,10]:

Mass conservation equation:

$$\frac{\partial \rho}{\partial t} = \nabla \cdot (\rho u) \tag{1}$$

where ρ is the density, t is the time, $\boldsymbol{u}$ is the velocity vector, and ∇ is the Hamiltonian operator.

Momentum conservation equation:

$$\rho\left[\frac{\partial u}{\partial t} + (u \cdot \nabla)u\right] + \nabla \cdot p = \rho g + f + \nabla \cdot \tau \tag{2}$$

where g is the acceleration of gravity, f is the volume force vector, τ is the viscous tension per unit area, and p is the pressure.

Energy conservation equation:

$$\frac{\partial}{\partial t}(\rho h) + \nabla[\rho h u] = \frac{\partial p}{\partial t} + u \cdot \nabla p - \nabla \cdot q_r + \nabla \cdot (k\nabla T) + \sum_i \nabla \cdot (h_i \rho D_i \nabla Y_i) \tag{3}$$

where h is the specific enthalpy, q_r is the thermal radiation flux, T is the temperature, k is the heat conductivity coefficient, D_i is the diffusion coefficient of the ith ingredient, and Y_i is the mass fraction of the ith ingredient.

Ingredient conservation equation:

$$\frac{\partial}{\partial t}(\rho Y_i) + \nabla \cdot (\rho Y_i u) = \nabla \cdot (\rho D_i \nabla Y_i) + m_i{}^m \tag{4}$$

where $m_i{}^m$ is the mass production rate of the ith ingredient.

Ideal gas state equation:

$$p_0 = \rho TR \sum_i (Y_i / M_i) = \rho TR / M \tag{5}$$

where p_0 is the background pressure, R is the molar gas constant, M is the molecular weight of mixed gas, and i is the ith ingredient.

2.1.2. Model Establishment of Double-Deck Bridge

FDS is used to establish a double-deck bridge section. The shape and dimensions of the bridge refer to the Shiziyang bridge in Guangzhou, China. In order to simplify the model, the bridge model in this paper does not consider the influence of the girder on the temperature field. A 3D view of the model of the bridge is shown in Figure 1. In the bridge model, the longitudinal direction of the bridge is in the x-direction with a 48 m length, the transverse direction is in the y-direction with a 42 m length, and the vertical direction is in the z-direction with varying length.

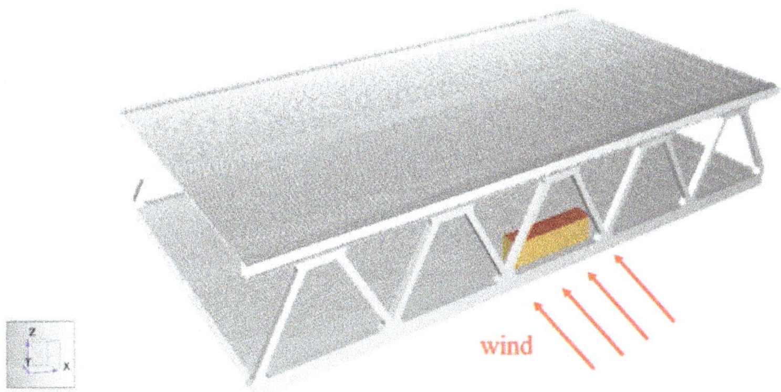

Figure 1. Three-dimensional (3D) view of the bridge.

The main material of the bridge is steel. The thermal parameters of the steel include thermal conductivity and specific heat. These parameters have been studied by many scholars and can be expressed as follows [11,12]:

$$\lambda(\theta) = \begin{cases} -0.022\theta + 48,\ 0 \le \theta \le 900\ ^\circ\mathrm{C} \\ 28.2,\ \theta > 900\ ^\circ\mathrm{C} \end{cases} \tag{6}$$

$$C(\theta) = 38.1 \times 10^{-8}\theta^2 + 20.1 \times 10^{-5}\theta + 0.473 \tag{7}$$

where λ is the thermal conductivity function of the steel, C is the specific heat function of the steel, θ is the temperature.

In this study, the thermal physical properties of steel at 20 °C are taken without considering the change of the material's thermal physical properties with temperature. The relevant parameters of the bridge and its properties are shown in Table 1.

Table 1. Parameters of materials used in the bridge model.

Property	Material	Thermal Conductivity (W/m^2 K)	Density (kg/m^3)	Specific Heat (kJ/kg K)
Bridge	Steel	47.56	7850	0.48

2.2. Parameter Setting

2.2.1. Model Parameters

The fire source is located at the outermost side of the bridge mid-span (as shown in Figure 1), and its size is 12 m × 3 m × 3 m. Since the study of bridge fire mainly focuses on an extreme fire source scale [13,14], the fire power is set as the heat release rate of the tanker during fire. Inagason [15] suggested the heat release rate for a tank fire be set at 200 MW. According to the results of the French regulations [16], the growth stage of a 200 MW fire source is 600 s, and the stable stage is 4200 s, so the simulation time is 4800 s in this work. The type of simulated fire source is a t^2-growth fire. The growth factor is calculated at the time specified by French regulations. It is appropriate to set the mesh size as 1/4–1/16 of the flame characteristic diameter (D^*) [17]. Employing the formula $D^* = (Q/\rho_0 C_p T_0 g^{1/2})^{2/5}$ [18], the flame characteristic diameter in this paper is calculated as 7.77 m, and thus, the appropriate mesh size is 0.49 m–1.94 m. Considering both simulation time and accuracy, 1 m is selected as the mesh size. The relevant parameters are shown in Table 2.

Table 2. Parameter settings of simulation.

Settings	Parameters
Ambient temperature	20 °C
Ambient pressure	101,300 Pa
Humidity	40%
Ventilation velocity	2 m/s
Bridge material	steel
Simulation time	4200 s
Fire type	t^2 unsteady state
Mesh size	1 m × 1 m × 1 m

The arrangement of thermocouples inside the utility tunnel is shown in Figure 2. The ceiling thermocouple is arranged 2 m apart, tiling the whole top deck of the bridge.

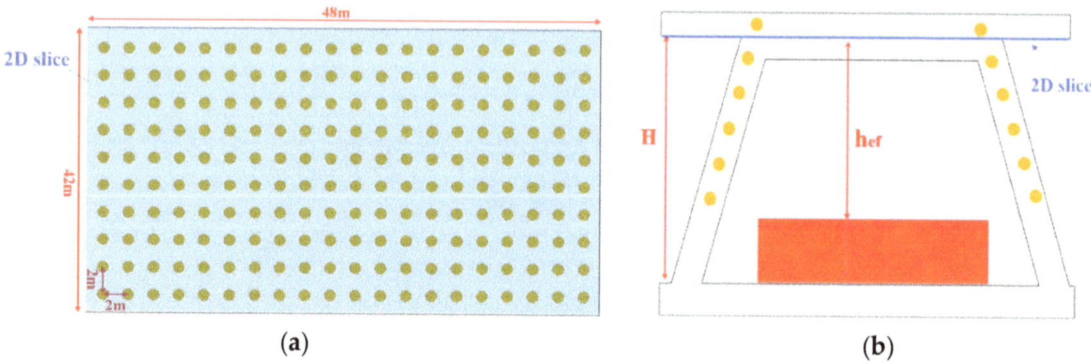

Figure 2. Thermocouple distribution in the truss and ceiling of double-deck bridge model: **(a)** bottom of ceiling; **(b)** inner surface of truss.

In order to ensure that the maximum excess temperature of the hot smoke layer below the ceiling can be measured, the thermocouples are arranged 0.05 m below the ceiling [19].

There are six thermocouples evenly arranged at the same intervals above the fire source on the inner surface of the truss (yellow points in Figure 2), and the thermocouples are arranged 0.05 m away from the surface of the truss. In addition, the 2D slice (blue graph in Figure 2) of the temperature is set at the bottom of the ceiling to observe the overall change of ceiling temperature. The height of the 2D slice is equal to the bridge deck spacing.

2.2.2. Operating Parameters

In the paper, the definition of bridge deck spacing is the vertical distance from the ceiling to the bottom deck of the double-deck bridge, as shown in Figure 2. Effective bridge deck spacing is the distance from the ceiling of the bridge to the surface of the fire resource. In this study, six different bridge deck spacings, i.e., 9.8 m, 10.6 m, 11.4 m, 12.2 m, 13.0 m, and 13.8 m, are selected and investigated, while other parameters remain unchanged. In addition, a ventilation velocity (2 m/s) is set according to the perennial wind speed in Guangzhou. As shown in Figure 1, the ventilation direction is along the y-direction.

3. Results and Discussion

3.1. *Influence Range of High-Temperature Smoke on the Ceiling*

Figure 3 shows the influence range of high-temperature smoke on the ceiling at different bridge deck spacings. As can be seen from the figure, the range of the high-temperature area gradually decreases. With the increase in bridge deck spacing, the area where the flame plume hits the ceiling decreases. When the temperature exceeds 300 °C, the properties of steel begin to decrease, so the 300 °C isotherm is selected as the influence range of the fire temperature field. The spread distance of fire influence decreases gradually along the x-direction. This is because the spread distance of flame plume along the ceiling becomes smaller as the bridge deck spacing increases. However, the spread distance of fire influence first remains unchanged and then decreases along the y-direction. This is because the flame plume tilts along the y-direction due to ventilation velocity.

3.2. *Maximum Excess Temperature Beneath the Ceiling*

In this paper, the maximum excess temperature beneath the ceiling is defined as the difference between the maximum temperature beneath the ceiling and the ambient temperature. The maximum excess temperature at different bridge deck spacings varies with time, as shown in Figure 4. It can be clearly seen that the maximum excess temperature changes with time and can be divided into three stages: slow growth stage, rapid growth stage, and relatively stable stage. In the slow growth stage, the maximum excess temperature increases slowly with time, which occurs between approximately 0 and 300 s. In the rapid growth stage, the maximum excess temperature increases rapidly with time, which occurs between approximately 300 and 600 s. In the relatively stable stage, the maximum excess temperature increases rapidly with time, which occurs between approximately 600 and 4200 s. The maximum excess temperature does not change significantly with time and fluctuates around a certain value. With the increase in bridge deck spacing, the overall excess temperature of the bridge ceiling decreases gradually. The reason Is that as the bridge deck spacing increases, the entrainment route of the air beneath the ceiling decreases, which increases heat flux loss.

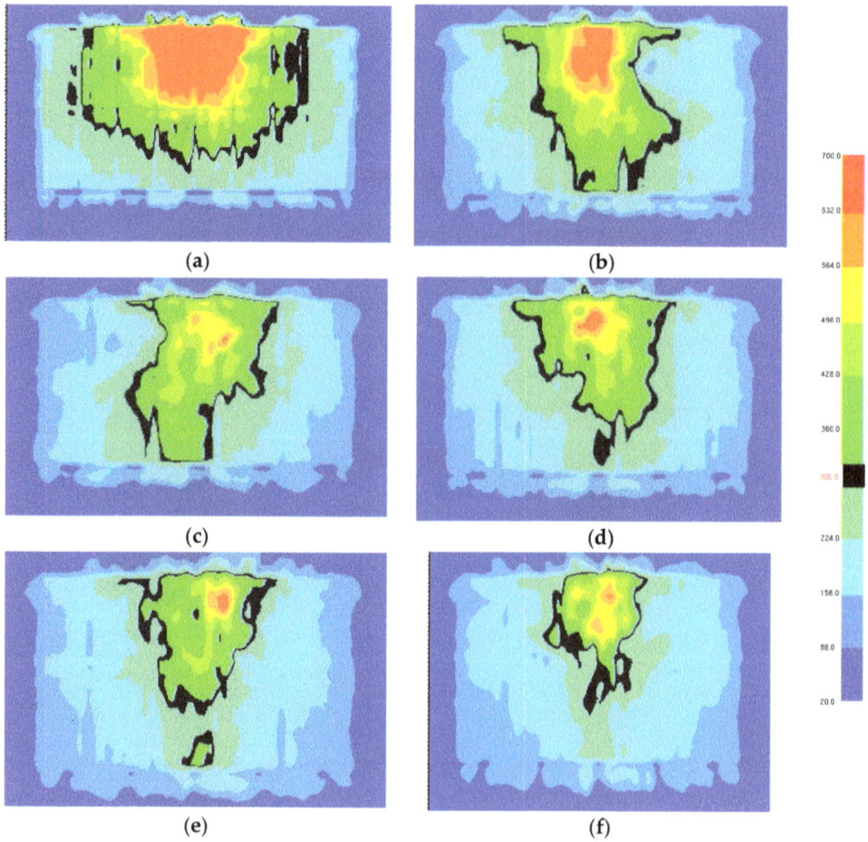

Figure 3. Influence range on the ceiling at different bridge deck spacings: (**a**) H = 9.8; (**b**) H = 10.6; (**c**) H = 11.4; (**d**) H = 12.2; (**e**) H = 13.0; (**f**) H = 13.8.

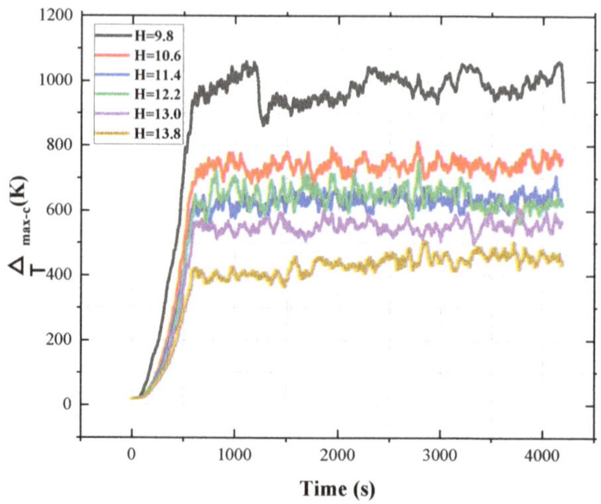

Figure 4. The maximum excess temperature versus time beneath the ceiling.

A lot of research has been performed on the formula of maximum excess tunnel ceiling temperature. Li proposed the formula of the maximum excess ceiling temperature under longitudinal ventilation velocity. The formula has been proved to be applicable to most cases by many scholars, which can be expressed as [20]:

$$\Delta T_{\max-c} = \frac{2.68C_T(1-\chi_r)g^{1/3}}{(\rho_0 C_p T_o)^{1/3}} \frac{Q}{uD^{1/3}{h_{ef}}^{5/3}}, u' > 0.19 \tag{8}$$

where $\Delta T_{max\text{-}c}$ is the maximum excess ceiling temperature, Q is the total heat release rate, T_o is the ambient temperature, C_p is the thermal capacity of air, C_T is coefficient, χ_r is the fraction of radiative heat release rate, g is the gravitational acceleration, D is the radius of the fire source, u is the ventilation velocity, and u' is the dimensionless ventilation velocity.

u' can be expressed as [21]:

$$u' = \frac{u}{w^*} \tag{9}$$

$$w^* = \left(\frac{Q_c g}{D\rho_0 C_p T_o}\right)^{1/3} \tag{10}$$

where w^* is the characteristic plume velocity, and Q_c is the convective heat release rate.

To verify whether the formula of the maximum excess temperature of the tunnel ceiling is suitable for a bridge under natural ventilation, Li's formula is used to fit the numerical simulation data. To reduce the error of numerical simulation results, the average value of the stable stage is selected as the maximum excess temperature beneath the ceiling. Figure 5 shows the fitting of the formula and simulation data. It is obvious that the formula of the maximum excess tunnel ceiling temperature also applies to the bridge. The simulation data of maximum excess temperature beneath the top floor can be correlated well with Equation (11):

$$\Delta T_{\max-c} = 0.397 \frac{Q}{uD^{1/3}{h_{ef}}^{5/3}} \tag{11}$$

where $\Delta T_{\text{max-c}}$ is the maximum excess temperature beneath the ceiling.

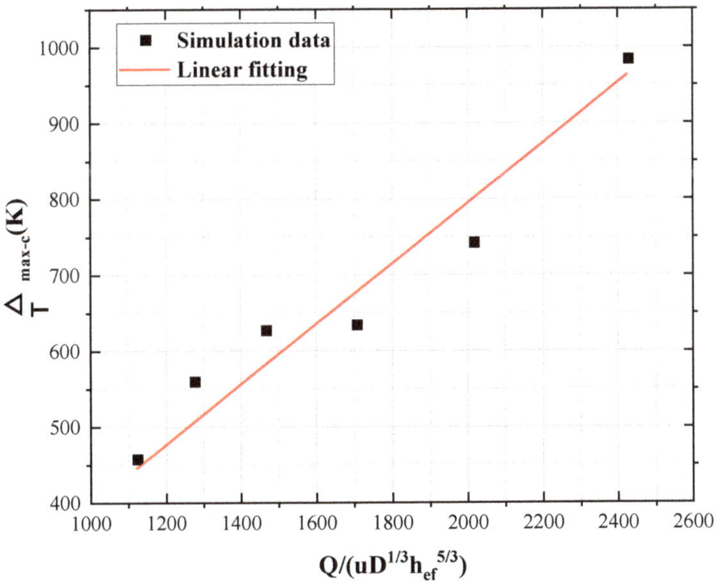

Figure 5. Fitting of Li's formula and simulation data.

Compared with Li's empirical formula, the coefficient is relatively small. The reason is as follows: firstly, the maximum excess temperature here is the average of the stable stage. Secondly, compared with tunnels, the double-deck bridge does not have sidewalls to limit heat dissipation.

3.3. Temperature Distribution of Truss

In this paper, the maximum excess temperature of the truss is defined as the difference between the maximum temperature of the truss and the ambient temperature. The maximum excess temperature of the truss at different bridge deck spacings varies with time and is shown in Figure 6. The maximum excess temperature varies with time and can also be divided into three stages: slow growth stage, rapid growth stage, and relatively stable stage. With the increase in the bridge deck spacing, the overall excess temperature gradually decreases. This is because as the bridge deck spacing increases, air entrainment increases and more cold air flows through the truss surface, which leads to the temperature decreasing. However, the excess temperature of the 12.2 m bridge deck spacing is bigger than that of the 11.4 m bridge deck spacing. This may be caused by the oxygen supply from outside of the 12.2 m bridge deck spacing becoming better than that of 11.4 m, which leads to the flame combustion reaction increasing.

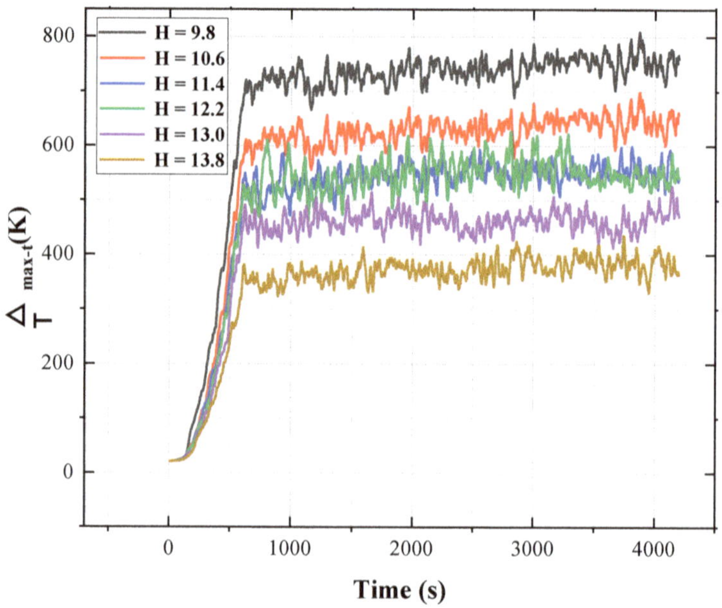

Figure 6. Maximum excess temperature of the truss versus time.

A great number of correlations for maximum excess temperature have been presented in the literature, even though there is some inconsistency with the different forms of them. An analysis of the maximum excess temperature in ventilation velocity has established the dependence of the dimensionless excess temperature ($\Delta T/\Delta T_{\text{max-c}}$) on the dimensionless HRR denoted as Q^* ($Q^* = Q/\rho_0 C_p T_0 g^{1/2} h_{ef}{}^{2/5}$ [22]) and the Froude number denoted as Fr ($Fr = u^2/gh_{ef}$ [22]). In addition, some scholars find that the HRR is closely related to the equivalent diameter of the fire source [9]. To quantify the maximum excess truss temperature, a dimensional analysis method is used. Based on the analysis above, the main factors affecting the maximum excess truss temperature are effective bridge deck spacing (h_{ef}), equivalent diameter of the fire source (D), flame heat release rate (Q), ambient air density (ρ_0), ambient temperature (T_0), specific heat capacity at constant pressure

(C_p), gravitational acceleration (g), and ventilation velocity (u). The maximum excess temperature of the truss can be expressed by the following formula:

$$f(h_{ef}, D, Q, \rho_0, C_p, g, u, T_0, \Delta T_{\max-t}) = 0 \tag{12}$$

where $\Delta T_{\text{max-}t}$ is the maximum excess temperature of the truss.

Based on the principle of dimensional consistency, the formula can be simplified as:

$$\frac{\Delta T_{\max-t}}{T_0} = \varphi\left(\frac{Q}{\rho_0 D^{\frac{7}{2}} g^{\frac{3}{2}}}, \frac{u}{D^{\frac{1}{2}} g^{\frac{1}{2}}}, \frac{C_p T_0}{Dg}, \frac{h_{ef}}{D}\right) \tag{13}$$

The right four terms of the formula can be combined:

$$\frac{\Delta T_{\max-t}}{T_0} = \varphi\left(\frac{Q}{\rho_0 D C_p T_0 u h_{ef}}\right) = \varphi(Q') \tag{14}$$

where Q' is the modified dimensionless heat release rate.

To reduce the error of the numerical simulation results, the average value of the stable stage is selected as the maximum excess temperature of the truss. Figure 7 shows that the simulation data of maximum excess temperature can be plotted as a function of $Q/\rho_0 D C_p T_0 u h_{ef}$ in this region. The simulation data of maximum excess temperature of the truss can be correlated well with Equation (15):

$$\frac{\Delta T_{\max-t}}{T_0} = 0.183\left(\frac{Q}{\rho_0 D C_p T_0 u h_{ef}}\right)^{1.257} \tag{15}$$

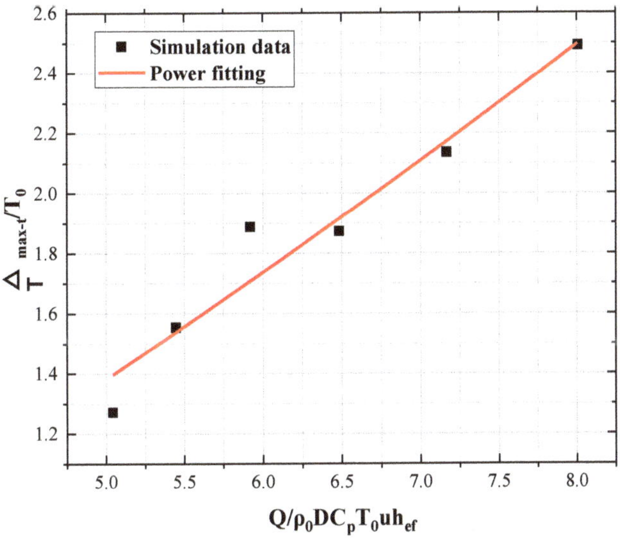

Figure 7. Dimensionless maximum excess temperature of the truss versus modified dimensionless heat release rate.

From one-dimensional analysis, Yang [23] estimated the heat loss intensity and established the energy equation of the gas beneath the ceiling. However, the excess temperature of the truss is mainly affected by the flame plume. Therefore, the energy diffusion of the micro length *dh* of the truss to the surrounding can be approximately expressed as:

$$q''Cdh = -c_p \dot{m} d(\Delta T) \tag{16}$$

where C is the perimeter of the contact surface between the element length and the surrounding environment, q'' is the heat flux to truss, and $\dot{m}$ is the mass flow rate of the truss.

q'' can be expressed by the following formula:

$$q'' = h_c \Delta T \tag{17}$$

where h_c is the lumped heat transfer coefficient.

$\dot{m}$ can be approximated as the mass rate of the micro side enters the flame plume:

$$\dot{m} = \rho_0 w_e h C \tag{18}$$

where w_e is horizontal entrainment velocity, which can be expressed as:

$$w_e = 1.94\alpha\left(\frac{g}{\pi C p T_0 \rho_0}\right)^{1/3} Q^{1/3} h^{-1/3} \tag{19}$$

Incorporating Equations (17)–(19) into Equations (16) and (19) can be obtained as follows:

$$\frac{d(\Delta T)}{\Delta T} = -\frac{h_c}{1.94\alpha\rho_0\left(\frac{g}{\pi C p T_0 \rho_0}\right)^{1/3} Q^{1/3}} \frac{dh}{h^{2/3}} \tag{20}$$

The boundary conditions are given by the following equation:

$$\begin{cases} h = h_0 \\ \Delta T = \Delta T_{\max - t} \end{cases} \tag{21}$$

where h_0 is the location of maximum excess temperature of the truss.

Incorporating Equations (20) and (21) can obtain the attenuation formula of excess temperature of the truss:

$$f(h) = \frac{\Delta T}{\Delta T_{\max - t}} = \mathrm{e}^{-A(h^{1/3} - h_0{}^{1/3})}, h > h_0 \tag{22}$$

where $A = \frac{3h_c}{1.94\alpha\rho_0\left(\frac{g}{\pi C p T_0 \rho_0}\right)^{1/3} Q^{1/3}}$.

Assuming that the excess temperature at the same distance from the maximum excess temperature position is almost same, the formula of the increasing excess temperature of the truss can be obtained as follows:

$$f(h) = f(2h_0 - h) \tag{23}$$

By substituting Equation (23) into Equation (22), Equation (24) can be expressed as:

$$f(h) = \frac{\Delta T}{\Delta T_{\max - t}} = \mathrm{e}^{-A\prime((2h_0 - h)^{1/3} - h_0{}^{1/3})}, h \le h_0 \tag{24}$$

To obtain the dimensionless temperature as a function of the dimensionless distance, h_{ef} is used as the characteristic scale to the dimensionless distance. Incorporating Equations (22) and (24), the dimensionless excess temperature above and below the position of the maximum temperature can be expressed as:

$$\frac{\Delta T}{\Delta T_{\max - t}} = \begin{cases} e^{-k\left(\left(\frac{h}{h_{ef}}\right)^{1/3} - \left(\frac{h_0}{h_{ef}}\right)^{1/3}\right)}, h > h_0 \\ e^{-k'\left(\left(\frac{2h_0 - h}{h_{ef}}\right)^{1/3} - \left(\frac{h_0}{h_{ef}}\right)^{1/3}\right)}, h \le h_0 \end{cases} \tag{25}$$

where k and k' are coefficients.

In this work, the maximum excess temperature of the truss is located at the top of the truss, namely $h_0 = h_{ef}$. So, the excess temperature of the truss can be expressed as:

$$\frac{\Delta T}{\Delta T_{\max-t}} = e^{-k'((2-\frac{h}{hef})^{1/3}-1)}, h \leq h_{ef} \tag{26}$$

Figure 8 shows that the simulation data fit well with the formula. The values of k' under different working conditions are shown in Table 3. When H $\leq$ 10.6, the value of k' is close to 1. However, when H > 11.4, the value of k' is larger than 1. This means that when the bridge deck spacing is less than 11.2, the excess temperature varies slightly along the truss. When the bridge deck spacing continues to increase, the change of excess temperature is great.

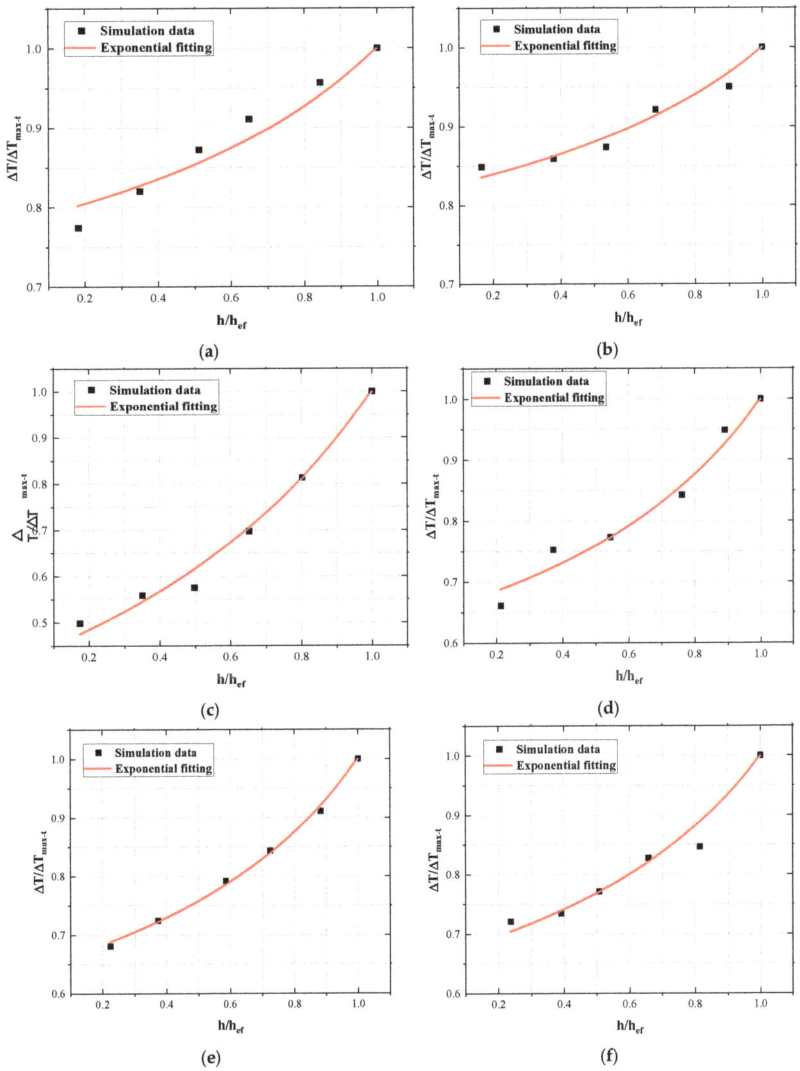

Figure 8. Excess temperature along the truss versus normalized effective height: (**a**) H = 9.8; (**b**) H = 10.6; (**c**) H = 11.4; (**d**) H = 12.2; (**e**) H = 13.0; (**f**) H = 13.8.

Table 3. Values of k' under different bridge deck spacings.

	H = 9.8	H = 10.6	H = 11.4	H = 12.2	H = 13.0	H = 13.8
k'	1.06	0.84	4.65	4.43	1.872	1.77

4. Conclusions

In this paper, the numerical simulation method is used to simulate the distribution characteristics of the smoke temperature field of a double-deck bridge during tanker fire under natural ventilation. The influence of bridge deck spacing on the smoke temperature field change is investigated. The following conclusions can be obtained:

1. For the different bridge deck spacings, the influence range of high-temperature smoke gradually decreases with the increase in bridge deck spacing. As the bridge deck spacing increases, the spread distance of the fire influence decreases gradually along the x-direction, where it first remains unchanged, then decreases along the y-direction.
2. The maximum excess temperature beneath the ceiling decreases with the increase in the bridge deck spacing. The maximum excess temperature function of the tunnel ceiling is also applicable to the bridge, but the coefficient is smaller than that of the tunnel experimental formula.
3. The excess temperature of the truss varies with time and can be divided into three stages: slow growth stage, rapid growth stage, and relatively stable stage. With the increase in the bridge deck spacing, the excess temperature gradually decreases. Through dimensionless analysis and simulation data fitting, an empirical formula is established, which indicates the dimensionless maximum excess temperature of the truss shows a power function growth trend with the increase in the modified dimensionless heat release rate.
4. The vertical excess temperature distribution of the truss above the fire source is investigated. The excess temperature increases along the truss, and the maximum excess temperature appears at the top of the truss. A model is established, which indicates the excess temperature along the truss conforms to exponential growth with the vertical distance under different bridge deck spacings (H). When $H \leq 10.6$ m, the excess temperature varies slightly along the truss. When $H \geq 11.4$ m, the change of excess temperature is great.

Author Contributions: Conceptualization, W.A.; Formal analysis, W.A.; Investigation, L.S., H.W. and T.Z.; Methodology, L.S. and T.Z.; Writing—original draft preparation, W.A. and L.S.; Writing—review and editing, W.A., H.W. and T.Z. All authors have read and agreed to the published version of the manuscript.

Funding: This research was funded by the Fundamental Research Funds for the Central Universities (Grant Number: 2021ZDPYYQ004).

Institutional Review Board Statement: Not applicable.

Informed Consent Statement: Not applicable.

Data Availability Statement: Not applicable.

Conflicts of Interest: The authors declare no conflict of interest.

References

1. Garlock, M.; Paya-Zaforteza, I.; Kodur, V.; Gu, L. Fire hazard in bridges: Review, assessment and repair strategies. *Eng. Struct.* **2012**, *35*, 89–98. [CrossRef]
2. Peris-Sayol, G.; Paya-Zaforteza, I.; Balasch-Parisi, S.; Alós-Moya, J. Detailed Analysis of the Causes of Bridge Fires and Their Associated Damage Levels. *J. Perform. Constr. Facil.* **2017**, *31*, 9. [CrossRef]
3. Mendes, P.A.; Valente, J.C.; Branco, F.A. Simulation of ship fire under Vasco da Gama Bridge. *ACI Struct. J.* **2000**, *97*, 285–290.
4. Branco, F.; Mendes, P.; Guerreiro, L. Special Studies for Vasco da Gama Bridge. *J. Bridge Eng.* **2000**, *5*, 3. [CrossRef]
5. Bennetts, I.; Moinuddin, K. Evaluation of the Impact of Potential Fire Scenarios on Structural Elements of a Cable-Stayed Bridge. *J. Fire Prot. Eng.* **2009**, *19*, 85–106. [CrossRef]

6. Smith, R.; Gogus, A.; Cormie, D.; Tehrani, P. Threat Mitigation Options in the Design of Cable-Stayed Bridges. In Proceedings of the Structures Congress 2014, Boston, MA, USA, 3–5 April 2014.
7. Ataei, H.; Mamaghani, M.; Aboutaha, R.S. Finite Element Analysis of Cable-Stayed Strands' Failure Due to Fire. In *Proceedings of the 7th Congress on Forensic Engineering, Miami, FL, USA, 15–18 November 2015*; American Society of Civil Engineers: New York, NY, USA, 2015.
8. Floyd, J.E.; Mcgrattan, K.B.; Hostikka, S.; Baum, H.R. CFD fire simulation using mixture fraction combustion and finite volume radiative heat transfer. *J. Fire Prot. Eng.* **2003**, *13*, 11–36. [CrossRef]
9. Karlsson, B.; Quintiere, J.G. *Enclosure Fire Dynamics*; CRC press: Boca Raton, FL, USA, 1999.
10. Mcgrattan, K.B.; Forney, G.P.; Floyd, J.; Hostikka, S.; Prasad, K. Fire Dynamics Simulator (Version 5): User's Guide; National Institute of Standards and Technology NIST. NIST Special Publication. 2007. Available online: https://nvlpubs.nist.gov/nistpubs/Legacy/SP/nistspecialpublication1019-5.pdf (accessed on 9 August 2022).
11. Lie, T.T. Fire Resistance of Circular Steel Columns Filled with Bar-Reinforced Concrete. *J. Struct. Eng.* **1994**, *120*, 1489–1509. [CrossRef]
12. Lie, T.T.; Celikkol, B. Method to calculate the fire resistance of circular reinforced concrete columns. *ACI Mater. J.* **1991**, *88*, 84–91.
13. Wang, Y.; Liu, M. Buckling Instability Behavior of Steel Bridge under Fire Hazard. *Math. Probl. Eng.* **2016**, *2016*, 11. [CrossRef]
14. Cui, C.; Chen, A.; Ma, R. Stability assessment of a suspension bridge considering the tanker fire nearby steel-pylon. *J. Constr. Steel Res.* **2020**, *172*, 106186. [CrossRef]
15. Ingason, H. Design fire curves for tunnels. *Fire Saf. J.* **2009**, *44*, 259–265. [CrossRef]
16. Gillard, J. Developments for the 21st Century. In Proceedings of the 9th International Symposium on Aerodynamics and Ventilation of Vehicle Tunnels, Aosta Valley, Italy, 6–8 October 1997.
17. Mcgrattan, K.B. *Fire Dynamics Simulator (Version 4): User's Guide*; National Institute of Standards and Technology: Gaithersburg, MD, USA, 2004.
18. Mcgrattan, K.B.; Baum, H.R.; Rehm, R.G. Large eddy simulations of smoke movement. *Fire Saf. J.* **1998**, *30*, 161–178. [CrossRef]
19. Tang, F.; He, Q.; Chen, L.; Li, P. Experimental study on maximum smoke temperature beneath the ceiling induced by carriage fire in a tunnel with ceiling smoke extraction. *Sustain. Cities Soc.* **2019**, *44*, 40–45. [CrossRef]
20. Li, Y.Z.; Lei, B.; Ingason, H. The maximum temperature of buoyancy-driven smoke flow beneath the ceiling in tunnel fires. *Fire Saf. J.* **2011**, *46*, 204–210. [CrossRef]
21. Quintiere, J.G.; Rinkinen, W.J.; Jones, W.W. The Effect of Room Openings on Fire Plume Entrainment. *Combust. Sci. Technol.* **1981**, *26*, 193–201. [CrossRef]
22. Kurioka, H.; Oka, Y.; Satoh, H.; Sugawa, O. Fire properties in near field of square fire source with longitudinal ventilation in tunnels. *Fire Saf. J.* **2003**, *38*, 319–340. [CrossRef]
23. Yang, D.; Huo, R.; Zhang, X.L.; Zhu, S.; Zhao, X.Y. Comparative study on carbon monoxide stratification and thermal stratification in a horizontal channel fire. *Build. Environ.* **2012**, *49*, 1–8. [CrossRef]

fire

MDPI

Article

Application Development of Smoke Leakage Test Apparatus for Door Sets in the Field

Hsuan-Yu Hung [1,*], Ching-Yuan Lin [1], Ying-Ji Chuang [1] and Chung-Pi Luan [2]

[1] Department of Architecture, National Taiwan University of Science and Technology, Taipei 10607, Taiwan; linyuan@mail.ntust.edu.tw (C.-Y.L.); chuang@mail.ntust.edu.tw (Y.-J.C.)
[2] Construction and Planning Agency, Ministry of The Interior, Taipei 10556, Taiwan; pierre@cpami.gov.tw
* Correspondence: D10913008@mail.ntust.edu.tw or auuv2000@gmail.com

Abstract: Heavy smoke from building fires is the main cause of casualties. As smoke typically diffuses through building openings, smoke control performance of building openings is critical to survival and requires considerable attention. In the past, the detection method could only be used in the laboratory, and the detection equipment could not be moved. Therefore, the main purpose of this research was to develop a methodology for field testing of smoke control properties of doors in order to ensure that the smoke control performance of doors tested in the laboratory and doors installed in the field can be realized without any discrepancy. Furthermore, this test method underwent a comparison test with the CNS 15038 "Method of Test for Evaluating Smoke Control Performance of Doors" for the same subject. The test results showed no significant difference based on independent sample testing, demonstrating the feasibility of this test method and test apparatus. The instrument developed by this research is light and easy to carry, and the operation method is simple. Such a test method can be applied to different doors and is non-destructive, non-hazardous, and reusable. In the future, by extending the design principle of the system, this test method can be applied to other fire protection equipment for the inspection of smoke control capabilities and can be used as a reference for relevant organizations to establish test specifications and standards.

Keywords: door; smoke; pressure difference; leakage test; building security

Citation: Hung, H.-Y.; Lin, C.-Y.; Chuang, Y.-J.; Luan, C.-P. Application Development of Smoke Leakage Test Apparatus for Door Sets in the Field. *Fire* **2022**, *5*, 12. https://doi.org/10.3390/fire5010012

Academic Editors: Chuangang Fan and Dahai Qi

Received: 10 December 2021
Accepted: 15 January 2022
Published: 18 January 2022

1. Introduction

According to previous research regarding building fires, the main factor contributing to human casualties is the dispersion of smoke through building openings [1]. The mortality rate due to this is over 50% greater than the mortality rate due to causes directly related to fires [2], which is why the smoke control performance of building openings has particular importance. Currently, the smoke control performance of building doors has been specified in various countries [3–9], requiring qualified smoke control doors to provide smoke control capabilities under both medium-temperature and ambient-temperature conditions. As a consequence of the complexity of fire scenarios, ambient-temperature smoke may also be life threatening [10]. For example, if a source of fire is relatively distant from a door, the temperature of the smoke near the door may be relatively low or even comparable to ambient temperatures. In such a scenario, while the door may not be damaged by the heat of the fire, individuals on the rear side of the door may be fatally exposed to excessive smoke inhalation. Therefore, a door's fire protection effectiveness may not be the primary concern. Instead, what is necessary is the door's capability of resisting against smoke inflow at ambient temperatures and under certain pressures. One example is BS 476-31 [11], which only requires smoke control performance at ambient temperature conditions. Doors with better smoke control capabilities at ambient temperatures can also prevent other toxic gases from entering the interior space under different circumstances, reducing the occurrence of related disasters.

According to Chuang et al. [12,13], Wu et al. [14], Tsai et al. [15], Gross [16], and Kuo et al. [17], under the same differential pressure conditions, when the temperature of the air increases, the volume increases and its density relatively decreases. As a result, the volume leakage of the air through the gap of the door will be reduced. In other words, if the volume of air leakage through the test body gap at ambient temperatures is available, identification whether the smoke control performance of the test subject at medium temperatures can meet the requirements of national regulations can be achieved, provided there is no deformation or damage of the material at medium temperatures. However, the components of doors for architectural use not only consist of steel, but also wood and organic materials, aluminum alloys, or plastics, as well as the use of filler strips, expansion materials, and glass windows. In 2000, Rakic [18] studied the effects of the presence and absence of filler strips on the smoke control performance of doors by means of UL 1784 [9]. The conclusion is that the installation of filler strips in doors could effectively increase their smoke control performance, demonstrating that filler strips and the methods used for the construction of door seams are the key to smoke control performance. However, these materials may be affected by thermal and physical or chemical transformations at medium temperatures, which may lead to inflation, flaming, or glass breakage. Therefore, medium-temperature testing is essential as this step of testing reveals the conditions of heat-induced deformation and damage to materials. To pass the test, the materials must possess capability of withstanding medium temperatures. Therefore, the premise of this field test method was that the test door should first pass the CNS 15038 [3] or ISO 5925-1 test [5] in the laboratory. The field test should only be conducted at ambient temperatures to determine whether the materials and methods applied can meet practical smoke control performance, so that the smoke control performance of doors tested in the laboratory and doors installed in the field can be realized without any discrepancy. In summary, we have developed a test method for the smoke control capabilities of door sets in the field through testing and theoretical analysis. In the past, the detection method could only be used in the laboratory, and the detection equipment could not be moved. The instrument developed by this research is light and easy to carry, and the operation method is simple. Such a test method can be applied to any doors and is non-destructive, non-hazardous, and reusable, enabling an immediate understanding of the door's smoke control performance in the field. In the future, by extending the design principle of the system, the test method can also be applied to other fire protection equipment for the inspection of smoke control capabilities and can be used as a reference for relevant organizations to establish test specifications and standards. This study offers the designs and descriptions of the equipment utilized with the intention of sharing this information for future reference.

2. Experimental Plan

2.1. Scope and Conditions

2.1.1. Scope

This test method is suitable for all doors, including single doors, double doors, elevator doors, roll-up doors, etc. It is a verification method for evaluating the smoke control performance of doors at ambient temperature. The prerequisite is that each door has passed the test of CNS 15038 [3].

2.1.2. Conditions

Before the field test, the thermometer, relative humidity meter, atmospheric pressure meter, gas volume flow meter, and differential pressure meter required for the test must be calibrated by a TAF—(Taiwan Accreditation Foundation) accredited laboratory. It must also meet the requirements of Section 2.2.

2.1.3. Terms and Definitions

① Door Assembly

A combination of fixed structures (such as doorframes), doors, and attached hardware accessories.

② Ambient Temperature

The ambient temperature of this test method is (25 ± 15) °C.

③ Leakage

The volume of air through the door assembly under a pressure difference.

2.1.4. Leakage Regulations

The volume leakage under a pressure difference of 10 and 25 Pa should be measured separately, with the benchmark of 25 Pa pressure difference, which is converted to the volume leakage under standard conditions that should not be greater than 25 m^3/h. In addition, the volume leakage under a pressure difference of 10 Pa should be free of abnormalities.

2.2. Experiment Apparatus

Apart from thermal treatment systems, the concepts utilized in the test method of this study are consistent with CNS 15038 [3], ISO 5925-1 [5], and BS 476-31 [11]. It is mainly based on the use of gas volume flow meters in conjunction with differential pressure meters to measure the volume leakage of test bodies. The configuration of the test equipment is shown in Figure 1, which can be further classified into three main components.

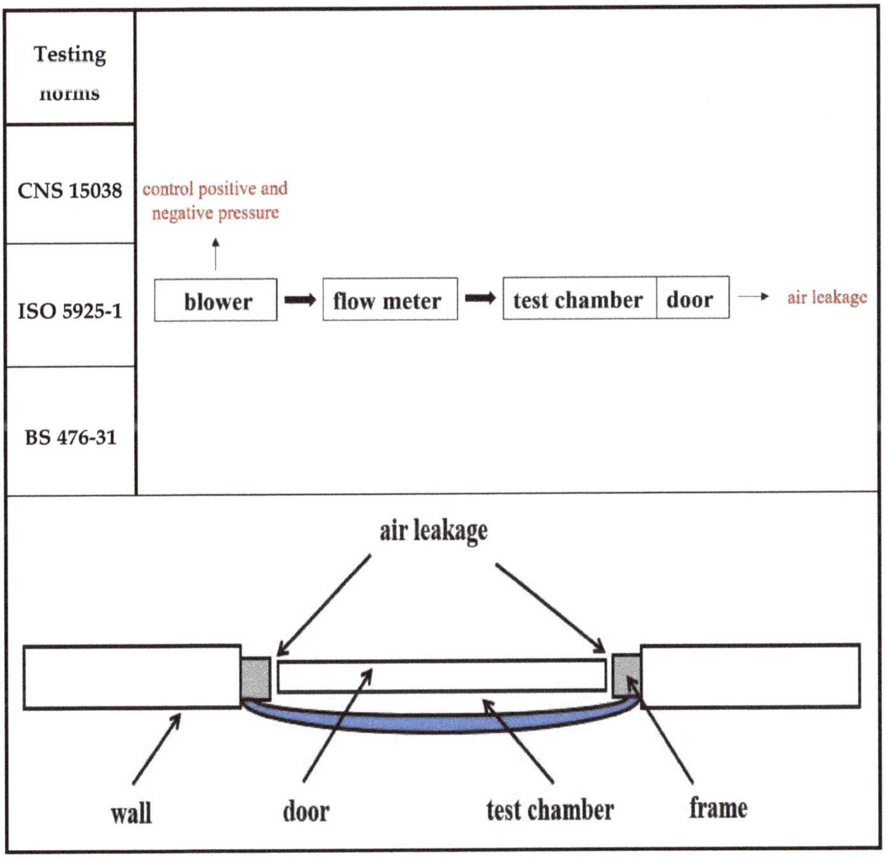

Figure 1. Configuration of the test equipment.

Component 1—Test chamber:

The test chamber consists of plastic sheeting and the test body, formed principally by anchoring the plastic sheeting around the door frame with airtight tape.

Component 2—Measurement system:

This includes ① a thermometer, ② a relative humidity meter, ③ an atmospheric pressure meter, ④ a gas volume flow meter, and ⑤ a differential pressure meter.

① Thermometer measurement range: −40 °C~+100 °C, with an accuracy of ±5%.
② Relative humidity meter measurement range: 0% RH~100% RH, with an accuracy of ±5%.
③ Atmospheric pressure meter measurement range: 300 hPa~1200 hPa, with an accuracy of ±5%.
④ Gas volume flow meter: Honeywell (Honeywell International, Inc., Charlotte, NC, USA) intelligent differential pressure transmitter with flow meter. The measurement range is 0 m^3/h ~75 m^3/h; the accuracy is ±2.5%; the applicable fluid temperature is −10 °C~+60 °C; and humidity is below 90%. The flow meter has an inlet and outlet pipe diameter of 50 mm and is installed between the blower outlet and the test chamber.
⑤ Differential pressure meter: Testo 510 (Testo SE & Co. KGaA, Titisee-Neustadt, Germany) pocket-type micro differential pressure meter, used to measure the difference in static pressure between the inside and outside of the test body; its measurement range is 0 hPa~100 hPa; its accuracy is ±0.03 hPa; and it is placed at 100 ± 10 cm from the surface of the test chamber in front.

Component 3—Inflating system:

The blowing engine from Y.H. Industrial Co., Ltd. (Taipei, Taiwan) with adjustable air speed can provide a stable air supply and pressure and is capable of maintaining a uniform pressure difference with a maximum air volume of 6.8 m3/min, 1/4 HP, a voltage of 220 V, three-phase electricity (Three-Phase), and a 50 mm outlet diameter, where the engine frequency control ranges from 0.01 HZ to 650.00 HZ in combination with a Teco Electric & Machinery Co. converter.

2.3. Test Principles

The configuration of the equipment used in the field test is shown in Figure 2, with a single test-chamber approach. The test chambers were uniformly positioned in the opening direction of the door (Figure 3) based on the previous results of Chuang et al. [19], which revealed that the volume of air leakage from both sides of the door at ambient temperatures measured was approximately equivalent. Therefore, it can be inferred that, regardless of which direction the door was tested, the results of leakage from the door should be representative. In addition, test chambers were uniformly established in the opening direction of the door for testing, in order to test the functionality of the test bodies through a functional operation of opening and closing each door more than five times [3] prior to starting the test. In the case where the test apparatus had been installed on site, as well as considering the use of positive pressure on the opening direction of the door according to common sense, there is no door frame to support the door, leaving only the automatic door closer and the lock. Therefore, the testing direction of the smoke control field test should be standardized to the open side of the door. This method of inspection should be relatively critical, and the actual volume of leakage after the installation of the door should be more detectable. The functionality of the automatic door closer and the lock can also be tested.

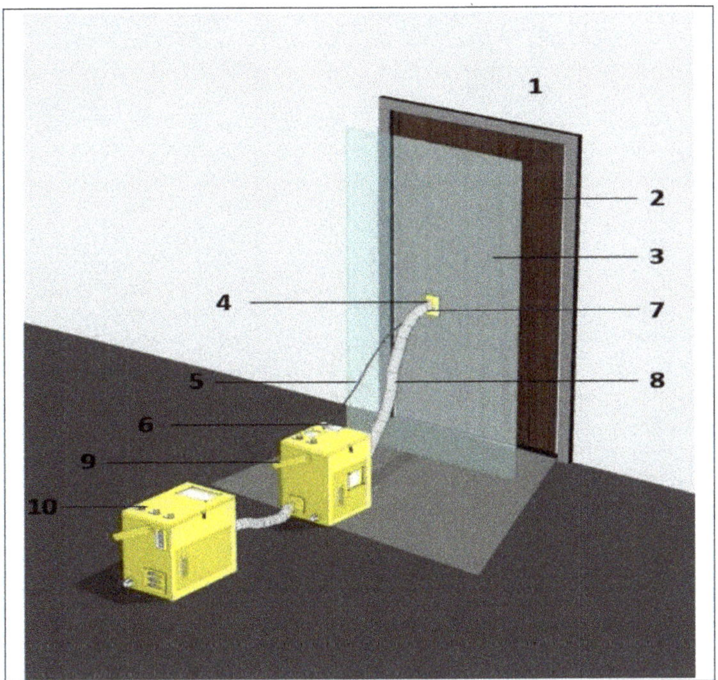

Items	Contents
Specimens	2. door
Test chamber	1. wall
	3. Plastic cloth
Equipment and pipeline	4. pressure port
	5. differential pressure pipe
	6. manometer
	7. tuyere
	8. air pipe
	9. flow meter
	10. industrial blower

Figure 2. Schematic diagram of the configuration of the equipment for the field test.

This test method was based upon the test chamber design example in Section 3.1 of CNS 15038 [3], the test system composition sketch in ISO 5925-1 [5], and the instrument assembly diagram in the journal published by Liu et al. [20]. In the previous version of CNS 15038 [21], amendments were made in accordance with the previous version of ISO 5925-1 [22] and ISO 5925-2 [23]. The subsequent revisions were also made in the updated version of CNS 15038 [3] in accordance with the updated version of ISO 5925-1 [5] and ISO 5925-2 [6], respectively. Regardless of how the specifications were revised throughout the years [3,5,6,21–23], it was found that with the exception of the medium-temperatures section, which is subject to a change in heating rate, all versions adopted a single test-chamber approach. Therefore, it was considered feasible for this study to utilize a single-chamber approach to measure the volume leakage of the door at ambient temperatures.

Figure 3. Field test images.

2.4. Test Procedure

The standard test procedure for the volume leakage of doors is as illustrated below:

Step 1. Pre-test environmental assessment

Measure the atmospheric pressure, temperature, and relative humidity of the test site. The temperature should be 25 ± 15 °C and the relative humidity should be 40–90% before the test is conducted.

Step 2. Door functionality assessment

Each door should be opened and closed (to a minimum of 30 degrees) more than 5 times [3]. In general, the opening direction of the door is the test surface. In the case of circumstances not permitting this, the other side can be used instead if the test cannot be conducted in this direction.

Step 3. Trimming the plastic sheeting

After the plastic sheeting is trimmed to the appropriate size for the door, a hole should be dug in its center for the tuyere and the pressure port.

Step 4. Establishment of the test chamber

Wipe the door frame to be tested with a rag, cover with plastic sheeting, and fix to the outside of the door frame with airtight tape.

Step 5. Equipment installation

Install the air supply tube and differential pressure tube in the tuyere and pressure port, and connect to the gas volume flow meter and differential pressure meter.

Step 6. Power on the equipment

After powering on the test equipment, zero and confirm the differential pressure meter and the gas volume flow meter.

Step 7. Environmental testing

Environmental testing should be conducted continuously for more than 3 min before the measurement of the volume leakage of the test body.

Step 8. Testing and recording

Activate the pressurization system and record the volumetric flow rate. Conduct the leakage measurement in accordance with the setting of 10 and 25 Pa differential pressure between the inside and outside of the test chamber. Under different differential pressure conditions, record every 30 s and calculate the average leakage during the 2 min.

Step 9. Restoration of the field site

Upon test completion, remove all equipment and restore the site to its original state.

3. Results and Discussion

3.1. Confirmation of Device Function

The precision instruments applied for testing purposes mainly involve a gas volume flow meter, a differential pressure meter, and a blowing engine, and all three of them should be calibrated. Upon completion of calibration, the measured results should be theoretically accurate provided that the equipment is not subjected to impacts. However, since this equipment is used in field tests, it is inevitable that there will be vibrations due to the frequent transportation. Therefore, before each official test, it is necessary to confirm the accuracy of the instrument to maintain the credibility of the test results. As a result, this study designed a test tube to be used for verification prior to official testing (with an inner diameter of 15.24 cm, a length of 100 cm, and a 2 cm diameter round hole in the center of one end of the tube for differential pressure measurement, with the other end connected to an air supply tube). Compare the theoretically calculated volume leakage value with the volume flow value measured by the test tube to confirm the accuracy of the flow meter measurement results. According to theoretical and practical studies in fluid mechanics, the flow coefficient is generally between 0.6 and 0.7 [14]. The different flow coefficients directly affect the calculated value of the flow in the openings. The factors determining the flow coefficients are complex, as the flow coefficients may be influenced by obstructions at the openings, such as the manner in which smoke flows through the openings [24] or the influence of human positioning [25], as well as varying gap widths [13]. For the sake of discussion, the theoretical volume leakage values can be calculated by Bernoulli's equation [26], as shown in Equation (1), assuming a flow coefficient of 0.6 to 0.7, an air temperature of 25 °C, and pressure differences of 10, 25, and 50 Pa.

$$Q = C \times A\sqrt{\frac{2\Delta P}{\rho}} \quad (1)$$

Q: Volume flow rate of air flow through apertures (m^3/s);
C: Flow coefficient (C = 0.6~0.7);
A: Flow area or ventilated area (m^2/s);
ΔP Differential pressure between the two sides of the air flow course (Pa);
ρ: Density of air entering the flow course (kg/m^3), $\rho = \frac{352.8}{273+T_\infty}$;
T_∞: Air temperature (°C).

The calculated leakage values are shown in Table 1, where the volume leakage amounts were 2.79, 4.41, and 6.24 m^3/h for pressure differences of 10, 25, and 50 Pa, respectively, at a flow coefficient of 0.6; then, the volume leakage amounts were 3.33, 5.27, and 7.46 m^3/h for pressure differences of 10, 25, and 50 Pa, respectively, at a flow coefficient of 0.7. In the future, if the measured volume leakage value is within the theoretical range of 0.60 to

0.70 flow coefficient, the volume leakage measurement equipment can be considered to be good and ready to be applied to the volume leakage measurement of subsequent doors.

Table 1. Theoretically projected volume leakage values.

Differential Pressure / Flow Coefficient	10 Pa	25 Pa	50 Pa
0.60	2.79 m^3/h	4.41 m^3/h	6.24 m^3/h
0.70	3.33 m^3/h	5.27 m^3/h	7.46 m^3/h

3.2. Flow Test Value Judgment

Kuo et al. [17] compiled the basic volume leakage requirements of test chambers from various countries. For instance, the previous version of ISO 5925-1 [22] and ISO 5925-2 [23] required the basic volume leakage of test chambers to be less than 1 m^3/h, and the previous version of CNS 15038 [21] required the basic volume leakage of test chambers to be less than 2 m^3/h. DIN 18095-1 [7] and DIN 18095-2 [8] specified that at a differential pressure of 50 Pa, the basic volume leakage of the test chamber shall not exceed 5 m^3/h; BS 476-31 [11] specified that at a differential pressure of 50 Pa, the basic volume leakage of the test chamber shall not exceed 7 m^3/h, and the new CNS 15038 [3] specified that the basic leakage of the test chamber shall be less than 7 m^3/h. In addition, UL 1784 [9] and JIS A1516 [4] do not specify the requirements of the basic volume leakage of the test chamber. The actual volume leakage of the door test body is defined as the total volume leakage minus the basic volume leakage of the test chamber. From the above, it can be seen that the basic volume leakage of the test chamber does not theoretically affect the actual volume leakage of the door. This is a simple mathematical procedure in which by subtracting the "basic volume leakage of the test chamber" from the "sum of the volume leakage of the test chamber", the volume leakage of the door test body can be found [27], calculated by Equation (2) and converted to the actual volume leakage of the test body in a standard condition.

$$Q_a{}' = \frac{Q_a}{(T + 273.15)} \times \left[k \times (p_a + p_m) - 3.795 \times 10^{-3} \times M_w \times p_{H_2O}\right] \quad (2)$$

Q_t: The sum of the volume leakage of the test body and the basic volume leakage of the test chamber (m^3/h);
Q_b: Basic volume leakage of the test chamber (m^3/h);
Q_a: Actual volume leakage at temperature (T + 273.15) and pressure (Pa + Pm) of the test body (m^3/h), $Q_a = Q_t - Q_b$;
$Q_a{}'$: Actual volume leakage of the test body under standard conditions (m^3/h);
T: Air temperature (°C);
k: Constant (293.15/101,325) = 2.89×10^{-3};
p_a: Atmospheric pressure (Pa);
p_m: Increased value of pressure (Pa);
M_w: Relative humidity (%);
p_{H_2O}: Saturation vapor pressure (Pa).

This test method excludes the step of measuring the basic volume leakage of the test chamber due to the field operating environment, which is bound to raise questions regarding the impartiality of this method. However, the basic volume leakage of the test chamber is in the range of 0~0.5 m^3/h after repeated tests in the field by taping the contact surface of the plastic sheeting to the door or wall. Theoretically, this method should allow one to more easily achieve airtightness as compared to a chamber consisting of a complex heating system, steel plates, piping, welding, screws, and adhesive strips in a laboratory. Although it is possible to apply adhesives in the field to bring the basic volume leakage of the test chamber to a level closer to 0 m^3/h, the basic volume leakage of the test chamber in the field should, for the sake of conservativeness, be 0.5 m^3/h for subsequent calculations.

Under different environmental conditions, the measured volume leakage values in the field should be corrected from Equation (2) to Equation (3), as the correction coefficients for different environmental conditions may vary.

$$Q_a{}' = \frac{Q_t - 0.5}{(T + 273.15)} \times \left[k \times (p_a + p_m) - 3.795 \times 10^{-3} \times M_w \times p_{H_2O}\right] \quad (3)$$

CNS 15038 [3], ISO 5925-1 [5], DIN 18095-1 [7], and BS 476-31 [11] all specify temperatures in the range of 25 ± 15 °C, i.e., from 10 to 40 °C for ambient-temperatures testing. In CNS 15038 [3], the volume leakage under a pressure difference of 10, 25, and 50 Pa should be measured separately, with the benchmark of 25 Pa pressure difference, which is converted to the volume leakage under standard conditions that should not be greater than 25 m^3/h. In addition, the volume leakage under a pressure difference of 10 Pa and 50 Pa should be free of abnormalities. The field test in this study excluded the detection of 50 Pa differential pressure for two reasons, the first of which is based on a comprehensive fire combustion study [28]. The differential pressure in general fires ranges from 10 to 15 Pa, while the field test in this study considered 25 Pa differential pressure as the benchmark, which is already a stringent standard and sufficient to meet the pressure derived from fires. The second is that the test chamber created with plastic tape and plastic sheeting may not be able to sustain the force of 50 Pa differential pressure for prolonged periods of time, which may cause the tape to loosen from the door frame. Although this can be improved by using a greater amount of, and better, tape, the test duration, operating cost, and the cleanliness of the door test body (which affects tape adhesion) are taken into consideration. Under conditions that do not affect the test results, the operation of measuring 50 Pa of differential pressure was excluded from the test results, and the field test was conducted only for the volume leakage measurement of 10 Pa and 25 Pa of differential pressure.

3.3. *Test Method Verification*

For the purpose of verifying the measurement capability of the field test equipment, a steel double flat door (2475 mm (door height) × 2370 mm (door width) × 55 mm (door thickness)) was first tested for volume leakage by this test method (Figure 4), and subsequently the same door was installed in the laboratory [17] for the CNS 15038 [3] volume leakage measurement. The test results were then compared. The temperature was 31.5 °C, with an ambient humidity of 73%, an ambient atmospheric pressure of 100,880 Pa, and a saturation vapor pressure of 4626.6 Pa. Volume leakage measurements were measured 10 times at a differential pressure of 10 Pa and 25 Pa, and recorded at 30, 60, 90, and 120 s during each testing period. The results of each test were then converted into the actual volume leakage of the test body under standard conditions (Figure 5), followed by statistical independent sample testing using SPSS [29] software (Table 2) to analyze the relationship between the two test results.

3.3.1. Levene's Test

The results of the volume leakage were analyzed by F-testing, where the p-value = 0.722 > 0.05 for a differential pressure of 10 Pa, indicating no significant difference between the two variables. The p-value = 0.718 > 0.05 for a differential pressure of 25 Pa, indicating no significant difference between the two variables.

3.3.2. Independent Samples *t*-Test

The F-test recognized that there was no significant difference in the variance of the volume leakage between the two. Therefore, in the t-test, under the condition of equal variances, the calculated t-statistic value of 10 Pa was −1.077, with a two-tailed p-value = 0.296 > 0.05. The calculated t-statistic value of 25 Pa was −1.467, with a two-tailed p-value = 0.160 > 0.05; therefore, the null hypothesis cannot be rejected.

Figure 4. Smoke control field test method verification.

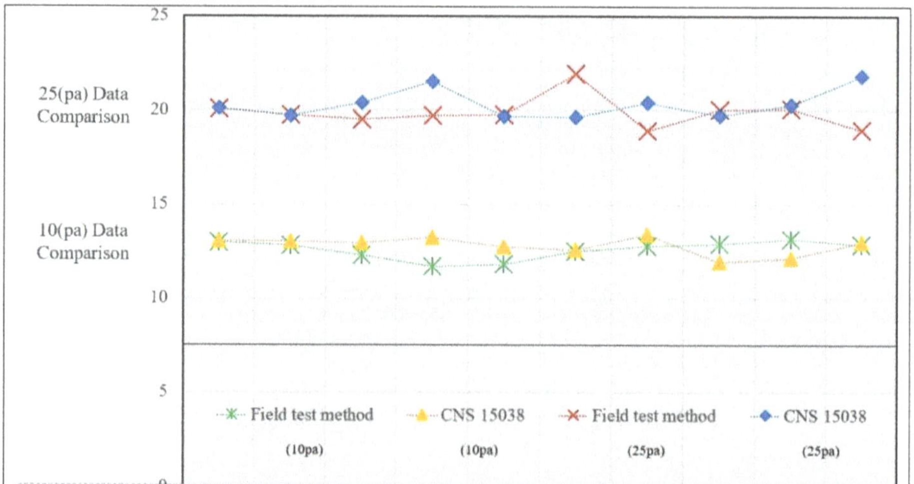

Experimental Data	1	2	3	4	5	6	7	8	9	10
Field test method	12.98	12.84	12.29	11.69	11.84	12.49	12.77	12.89	13.14	12.87
CNS 15038	13.07	13.04	12.96	13.23	12.73	12.58	13.38	11.96	12.16	12.98
Field test method	20.09	19.76	19.51	19.72	19.75	21.93	18.92	20.04	20.09	18.95
CNS 15038	20.11	19.71	20.39	21.53	19.68	19.62	20.41	19.72	20.26	21.82

Figure 5. Smoke control field test method and CNS 15038 test results.

Table 2. Independent sample testing.

Differential Pressure	Testing Methodology	Levene's Test		Independent Samples *t*-Test					
		F	*p*-Value	t	Degree of Freedom	*p*-Value (Two-Tailed)	Mean Difference	SD Difference	CI 95%
10 Pa	Equal Variance	0.131	0.722	−1.077	18	0.296	−0.229	0.21257	Lower Limit −0.6756 Upper Limit 0.2176
	Unequal Variance			−1.077	17.895	0.296	−0.229	0.21257	Lower Limit −0.67579 Upper Limit 0.21779
25 Pa	Equal Variance	0.134	0.718	−1.467	18	0.160	−0.536	0.36530	Lower Limit −1.30346 Upper Limit 0.23146
	Unequal Variance			−1.467	17.958	0.160	−0.536	0.36530	Lower Limit −1.30359 Upper Limit 0.23159

3.3.3. Conclusions

The results of the independent sample testing and analysis showed no significant difference in the volume leakage test results between the two tests at ambient temperatures under differential pressures of 10 Pa and 25 Pa. This suggests that the data measured by the smoke control field test method and the instrumentation used in this study were accurate and reliable, and can therefore be used for field tests.

3.4. Actual Experiment

To investigate the feasibility of the smoke control field test method, the equipment was moved to a construction site to conduct tests on a total of 20 doors (test numbers: A to T). The test time is from summer (average 29 °C) to winter (average 15 °C) in Taiwan, to test whether the equipment still has the ability to measure in the face of temperature, air pressure, and humidity changes. During the testing process, the site environment varied for each test, including room doors in first-floor residences, room doors in long-term care facilities, room doors in nursing facilities, escape doors in tunnels, and safety ladder doors on the fortieth floor. The test doors consisted of wood, plastic, aluminum, and iron, respectively. In each test, the plastic sheeting can be fixed to the door only by tape to form the test chamber. The tape used in the smoke control field test was made of aluminum foil (thickness 0.12 mm, width 14 cm, tensile strength 1020 N/100 mm, elongation at break 10%, and adhesion 51 N/100 mm). The entire test process can be completed by two testing personnel within 30 min. The results of each test were converted into the actual volume leakage of the test body under standard conditions (Figure 6), which can be found to possess specific pattern properties at 10 Pa and 25 Pa. As an example, for the test result of test number A, the actual volume leakage under a pressure difference of 25 Pa was converted into 13.73 m^3/h under standard conditions, which can be calculated by Equation (1). The theoretical volume leakage value under 10 Pa pressure difference was 5.64 m^3/h, which was only 0.19 m^3/h away from the field test result of 5.83 m^3/h. With a discrepancy of less than 1 m^3/h between the theoretical volume leakage value derived from Bernoulli's equation [26] and the volume leakage value measured by the field test, the test data of this study have been proved to be reliable. Under a pressure difference of 25 Pa, the results of the D, H, O, and Q tests were 30.04 m^3/h, 39.68 m^3/h, 28.73 m^3/h, and 37.42 m^3/h. The test results were all larger than 25 m^3/h and not in compliance with the standard. It was noted that the reason for the failure could be attributed to the inability to adequately fill the airtight strips due to oversized door seams. In the C, I, and P tests, the gap between the door and the floor was 1, 2, and 1.9 cm. Although the air supply system was operated at full power, the door was installed with an automatic descending airtight strip that failed to adhere closely to the ground, resulting in an excessive volume of leakage from the test chamber that prevented the pressure in the test chamber from rising. Other doors that passed the test (A, B, E, F, G, J, K, L, M, N, R, S, and T) were observed to have airtight strips set up in accordance with regulations and installed in appropriate positions. The test results further corresponded to the fact that while the doors could pass the CNS 15038 [3] test standard in the laboratory, the doors installed in the field may not necessarily achieve the same performance as in the laboratory. Therefore, it is necessary to investigate the actual smoke control performance of the doors in the field through field tests to help maintain the safety of personnel during evacuation scenarios.

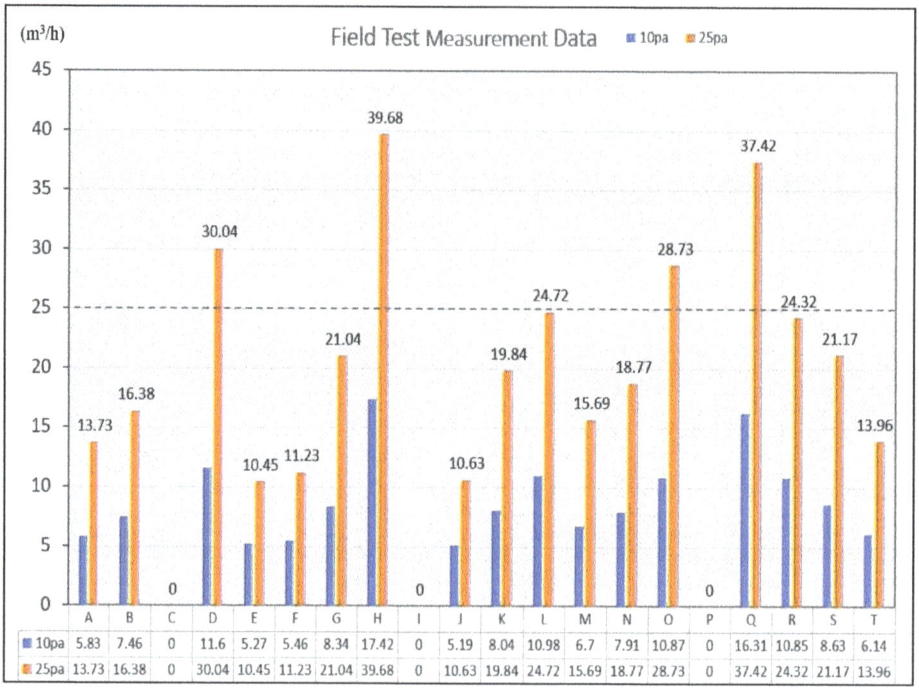

Figure 6. Field test measurement data.

4. Conclusions

A test tube was designed to allow for verification prior to official testing. Theoretically projected volume leakage values can be compared with the volume flow values measured by the test tube, thereby allowing for an immediate understanding of the accuracy of the measurement results prior to field testing.

Upon analysis, the general differential pressure in the fire scenarios varied from 10 to 15 Pa, considering test duration, operating costs, and the cleanliness conditions of the test bodies (which affects tape adhesion). The operation of measuring 50 Pa differential pressure was excluded under the condition that it does not affect the test results. Therefore, the field test was conducted only for differential pressures of 10 Pa and 25 Pa for the volume leakage measurement.

A comparison test was conducted for this test method and the CNS 15038 [3] method to evaluate the smoke control performance of doors for the same subject, and a statistically independent sample was examined using SPSS [29] software.

The whole test procedure can be completed by two testing personnel within 30 min, without damaging the door or contaminating the environment. It is demonstrated that this method can complete measurements in any environment, and the test results possess certain reliability.

Among the 20 doors tested in the field, only 13 doors passed the field test, thereby demonstrating that the doors installed in the field do not necessarily achieve the same smoke control performance as those in the laboratory, The smoke control performance may deteriorate due to increased gaps between doors and frames or wear and tear of door hinges over time. The above scenarios highlight the need to manage the smoke control performance from the user end in order to ensure the safety of personnel.

Upon many field experiments, the actual leakage is very close to the theoretical values. The volume leakage will increase with enlarged openings or increased pressure differences. In addition to the large leakage rate through the gap beneath the door, other volumetric airflow rates through lockset, hinges, top, and side gaps are very small.

This test method can be directly applied to various doors, including single and double doors, elevator doors, and roll-up doors. In the future, by extending the design principle of the system, the test method can be applied to other fire protection equipment for the inspection of smoke control capabilities, such as building ducts and fire escapes.

Author Contributions: Conceptualization, H.-Y.H., C.-Y.L., and Y.-J.C.; data curation, H.-Y.H.; formal analysis, C.-P.L.; Investigation, H.-Y.H.; methodology, H.-Y.H., C.-Y.L., Y.-J.C.; validation, H.-Y.H.; visualization, H.-Y.H.; writing—original draft, H.-Y.H.; writing—review and editing, H.-Y.H. All authors have read and agreed to the published version of the manuscript.

Funding: This research received no external funding.

Institutional Review Board Statement: Not applicable.

Informed Consent Statement: Not applicable.

Data Availability Statement: The data are contained within the article. The data presented in this study are available in Figures 5 and 6 and Table 2.

Conflicts of Interest: The authors declare no conflict of interest.

References

1. Lee, S.C.; Lin, C.Y.; Chuang, Y.J.; Lin, Y.S. A Computer-Based Simulation and Evaluation: Applying an Automatic Sprinkler System for Extinguishing Scooter Fires in Arcade Areas. *J. Eng. Res.* **2021**, *9*, 14–24. [CrossRef]
2. Clarke, F.B. Physiological Effects of Smoke: Managing Escape. *ASHRAE J.* **1997**, *39*, 47–56.
3. *CNS 15038*; Method of Test for Evaluating Smoke Control Performance of Doors. Bureau of Standard, Metrology and Inspection (BSMI): Taipei, Taiwan, 2009.
4. *JIS A1516*; Windows and Door Sets—Air Permeability Test. Japanese Industrial Standards: Tokyo, Japan, 1998.
5. *ISO 5925-1*; Fire Test-Evaluation of Performance of Smoke Control Door Assemblies-Part 1: Ambient Temperature Test. International Organization for Standardization (ISO): Geneva, Switzerland, 2007.
6. *ISO 5925-2*; Fire Test-Smoke Control Door and Shutter Assemblies-Part 2: Commentary on Test Method and Test Data Application. International Organization for Standardization (ISO): Geneva, Switzerland, 2006.
7. *DIN 18095-1*; Smoke Control Doors; Concepts and Requirements. German National Standard: Berlin, Germany, 1988.
8. *DIN 18095-2*; Smoke Control Doors; Type Testing for Durability and Leakage. German National Standard: Berlin, Germany, 1991.
9. *UL 1784*; Air Leakage Tests of Door Assemblies. Underwriters Laboratories Inc. (UL): Northbrook, IL, USA, 1990.
10. Lin, B.S.M.; Lin, C.Y.; Kung, C.W.; Lin, Y.J.; Chou, C.C.; Chuang, Y.J.; Hsiao, G.L.K. Wayfinding of Firefighters in Dark and Complex Environments. *Int. J. Environ. Res. Public Health* **2021**, *18*, 8014. [CrossRef] [PubMed]
11. *BS 476-31*; Methods for Measuring Smoke Penetration Through Door Sets and Shutter Assemblies. British Standards Institution (BSI): London, UK, 1983.
12. Chuang, Y.H.; Chuang, Y.J.; Lin, C.Y. Using a New Testing Method to Measure Smoke Leakage of Existing Doors. *J. Appl. Fire Sci.* **2006**, *16*, 21–33. [CrossRef]
13. Chuang, Y.J.; Luan, C.P.; Lin, C.Y.; Chen, P.H. Development and Application of Smoke Leakage Test Apparatus. *J. Fire Sci.* **2009**, *27*, 213–234. [CrossRef]
14. Wu, C.W.; Lin, T.H.; Chen, C.J.; Tsai, M.J. Smoke Leakage Through Wall Openings in a Fire. *Exp. Therm. Fluid Sci.* **2007**, *32*, 29–37. [CrossRef]

15. Tsai, T.H.; Kuo, S.Y.; Tseng, Y.T.; Tang, C.H.; Chuang, Y.J.; Lin, C.Y. Rates of Smoke Leakage Through Fire Stops. *J. Eng. Res.* **2013**, *1*, 231–250.
16. Gross, D. Estimating Air Leakage Through Doors for Smoke Control. *Fire Safety. J.* **1991**, *17*, 171–177. [CrossRef]
17. Kuo, S.Y.; Tseng, Y.T.; Chuang, Y.J. Comparison of Test Apparatus for Determining the Smoke Leakage Rate of Fire Doors. *J. Food Agric. Environ.* **2013**, *11*, 2831–2841.
18. Rakic, J. The Performance of Unit Entry Doors when Exposed to Simulated Sprinkler Controlled Fires. Fire Australia. 24–28 February 2000. Available online: citeseerx.ist.psu.edu/viewdoc/download?doi=10.1.1.527.1321&rep=rep1&type=pdf (accessed on 30 November 2021).
19. Chuang, Y.J.; Tsai, T.H.; Chuang, Y.-H.; Lin, C.Y.; Huang, C.H.; Chen, P.H. Performance Assessment of Single-Leaf Timber Door in a Smoke Leakage Test. *J. Appl. Fire Sci.* **2006**, *16*, 101–114. [CrossRef]
20. Liu, F.; Gong, N.; Yan, Z.J. Opening Flow Coefficient and Its Effect on Fire Smoke Flow. *J. Chongqing Jianzhu Univ.* **2000**, *22*, 86–92.
21. *CNS 15038*; Method of Test for Evaluating Smoke Control Performance of Doors. Bureau of Standard, Metrology and Inspection (BSMI): Taipei, Taiwan, 2002.
22. *ISO 5925-1*; Fire Test-Evaluation of Performance of Smoke Control Door Assemblies-Part 1: Ambient Temperature Test. International Organization for Standardization (ISO): Geneva, Switzerland, 1981.
23. *ISO 5925-2*; Fire Test-Smoke Control Door and Shutter Assemblies-Part 2: Commentary on Test Method and Test Data Application. International Organization for Standardization (ISO): Geneva, Switzerland, 1997.
24. Tamura, G.T. Fire Tower Tests on Vestibule Pressurization for Protection of Stair Shafts. *ASHRAE Trans.* **1994**, *100*, 981–989.
25. Tamura, G.T. Stair Pressurization Systems for Smoke. Control: Design Considerations. *ASHRAE Trans.* **1989**, *95*, 184–192.
26. Hewitt, P.G. Bernoulli's Principle Understanding Bernoulli's Principle as it Applies to Aerodynamic Lift. *Natl. Sci. Teach. Assoc.* **2004**, *71*, 51–55.
27. Chou, T.L.; Tang, C.H.; Chuang, Y.J.; Lin, C.Y. Study on Smoke Leakage Performance of Suspended Ceiling System. *Sustainability* **2020**, *12*, 7244. [CrossRef]
28. Klote, J.H. *Smoke Movement Through a Suspended Ceiling System*; NBSIR-81-2444; Center for Fire Research, National Bureau of Standards, NIST Interagency: Washington, DC, USA, 1982.
29. Allen, P.; Bennett, K.; Heritage, B. *SPSS Statistics Version 22: A Practical Guide*; Cengage Learning Australia: Victoria, Australia, 2014.

fire

Technical Note

Investigation of the Temperature Beneath Curved Tunnel Ceilings Induced by Fires with Natural Ventilation

Haowen Tao *, Zhisheng Xu and Dongmei Zhou

School of Civil Engineering, Central South University (CSU), Changsha 410075, China; zhshxu@csu.edu.cn (Z.X.); 204811140@csu.edu.cn (D.Z.)
* Correspondence: hw.tao@csu.edu.cn; Tel.: +86-731-8265-5177

Abstract: The distribution characteristics of the temperature below ceilings in curved tunnel fires have not been quantitatively studied. A small-scale tunnel was constructed in this work to study the maximum temperature and longitudinal attenuation of the temperature below ceilings induced by fires in a curved tunnel with natural ventilation. Different tunnel turning radiuses and fire heat release rates were taken into account. The results show that the distribution characteristics of temperature below the tunnel ceiling is hardly affected by the tunnel turning radius in a scenario where the flame plume impinges on the ceiling (strong-plume-driven). The fire-induced maximum temperature and longitudinal attenuation of temperature in curved tunnels are comparable to those of straight tunnels. Improved correlations are proposed for the condition of a strong-plume-driven ceiling jet, and the measured value of the temperature of the experiment collapsed well. This work may enhance the understanding of temperature distributions in curved tunnel fires.

Keywords: curved tunnel fire; naturally ventilated tunnel; maximum temperature; longitudinal temperature attenuation

Citation: Tao, H.; Xu, Z.; Zhou, D. Investigation of the Temperature Beneath Curved Tunnel Ceilings Induced by Fires with Natural Ventilation. *Fire* **2022**, *5*, 90. https://doi.org/10.3390/fire5040090

Academic Editors: Chuangang Fan and Dahai Qi

Received: 26 May 2022
Accepted: 22 June 2022
Published: 27 June 2022

1. Introduction

The construction of urban tunnels in metropolitan areas provides an effective solution for traffic congestion [1]. Curved tunnels are commonly used [2] as parts of urban tunnel systems (for example, large underground interconnected infrastructure, urban traffic link tunnels, etc.). Depending on the tunnel length, parts of the urban tunnels can be operated under natural ventilation. However, it appears that little research has been conducted on the temperatures below curved tunnel ceilings induced by fires under a natural ventilation scenario.

Maximum temperature rise and longitudinal temperature attenuation properties are two aspects of concern in a tunnel fire [3–5] and can serve as references for tunnel fire protection. In 1972, Alpert [6] presented a model for the prediction of maximum temperature rise induced by fires, given as:

$$\Delta T_m = 16.9\frac{\dot{Q}^{2/3}}{H_{ef}^{5/3}} \quad (1)$$

where ΔT_m is the maximum temperature rise below the ceiling (K), H_{ef} is the distance between the fire surface and the ceiling in the vertical direction (m), and $\dot{Q}$ is the fire heat release rate (kW).

In addition, in 1979 Heskestad and Delichatsios proposed the following equation for maximum temperature rise prediction in dimensionless form [7]:

$$\frac{\Delta T_m}{T_\infty} = 6.3\dot{Q}^{*2/3} \quad (2)$$

where $\dot{Q}^*$ can be calculated as:

$$\dot{Q}^* = \frac{\dot{Q}}{\rho_\infty T_\infty c_p \sqrt{g} H_{ef}^{5/2}}, \tag{3}$$

where T_∞, ρ_∞, and c_p are the temperature (K), density (kg/m^3), and specific heat at a constant pressure (kJ/(kg·K)) of ambient air, respectively, and g is the acceleration of gravity (m/s^2).

It should be noted that for a specified scenario where T_∞, ρ_∞, and c_p are constants expressions of Equations (1) and (2) will be similar.

As for the longitudinal temperature attenuation, Delichatsios [8] investigated the phenomenon of smoke spreading along a beamed ceiling in building fires, and proposed a correlation for the longitudinal temperature attenuation, given as:

$$\frac{\Delta T_x}{\Delta T_0}\left(\frac{l}{H}\right)^{1/3} = 0.49 e^{-6.67St\left[\frac{x}{H}\cdot\left(\frac{l}{H}\right)^{1/3}\right]}, \tag{4}$$

where ΔT_x is the temperature rise for a certain position denoted x (K), l is the half width between the beams (m), x is the distance between the fire source and the position x in the horizontal direction (m), H is the height of the ceiling (m), and St represents the Stanton number.

It is noted that both the models of the maximum temperature and the temperature attenuation were originally used for building fires. It was then proven by other scholars that the predicted model for tunnel fires was similar. Li et al. [9] proposed a correlation for the maximum temperature in tunnel fires. For scenarios with natural ventilation, the correlation is similar to Equation (1), and is given as:

$$\Delta T_m = 17.5 \frac{\dot{Q}^{2/3}}{H_{ef}^{5/3}}. \tag{5}$$

Meanwhilem Hu et al. [10] indicated that the temperature attenuation along the tunnel can be expressed using an exponential function, which is given as:

$$\frac{\Delta T_x}{\Delta T_0} = e^{-Kx} \tag{6}$$

where ΔT_0 is the temperature rise at the point of reference (K), and K is a coefficient.

In addition, it has been shown in previous studies that the temperature in a tunnel fire can be affected by various aspects, such as the tunnel geometry, inclination, location of fires, ventilation conditions, vehicle blockage, etc. [11–16].

Zhang et al. [13] revealed that the maximum temperature below the tunnel ceiling varies with the tunnel slope. Their results showed that an increase in the tunnel slope decreases the maximum temperature. Based on these findings, a correlation of maximum temperature induced by fires for a sloping tunnel was proposed. Tang et al. [15] indicated that the maximum temperature below tunnel ceilings is influenced by the variation of fire locations in the transverse direction through experimental studies. A dimensionless coefficient k was introduced to modify the effect of different fire-source locations. The effect of mechanical smoke extraction was also discussed [16], in which the characteristic smoke attenuation in the transverse direction was presented. Ingason [17] conducted a small-scale fire test to reveal the tunnel fire behaviors under longitudinal ventilation. It was shown that the ventilation velocity has little effect on the maximum temperature for a fire source with a full-scale heat release rate larger than 100 MW. Chen [18] conducted a small-scale experiment to study the temperature attenuation in longitudinally ventilated tunnels, in which the modified characteristic length was introduced to develop a predictive correlation. Tao et al. [19] experimentally investigated the temperature profile induced by fires in a tunnel with the centralized smoke exhaust on a single side. The influence of vehicle blockage on maximum temperature under longitudinal ventilation was investigated by Tang et al. [20], in which different distances between the fire source and the blockage were taken into account.

However, the former models of temperature distribution beneath the ceilings of tunnel fires were established based on straight tunnels. When fire occurs in a curved tunnel, the resistance of smoke flow spreading differs from that in a straight tunnel [21]. The results of former studies show that the curve of a tunnel leads to a different characteristic of smoke spread under longitudinally forced ventilation [22,23]. However, the effect of tunnel turning radius on the spread of fire smoke and temperature distribution in a naturally ventilated tunnel has not been considered.

In this study, an investigation is conducted via a small-scale curved tunnel experiment to disclose the temperature profiles induced by fires (both maximum temperature rise and longitudinal temperature attenuation) below tunnel ceilings. Improved correlations are provided that may be helpful for understanding smoke spreading characteristics in a curved tunnel.

2. Experimental Setup

2.1. Small-Scale Curved Tunnel

A small-scale curved tunnel is constructed based on the Froude scaling law [24,25], whose dimensions are 3 m in length, 0.22 m in height, and 0.32 m in width. Figure 1 shows the apparatus of the small-scale curved tunnel. Fire-proof glass with a thickness of 6 mm is adopted for the tunnel sidewalls and a fire-proof plate with a thickness of 3 mm is used for the tunnel ceiling.

Figure 1. Apparatus of small-scale curved tunnel.

Figure 2 shows the thermocouple arrangement in the small-scale curved tunnel. The tunnel longitudinal length and tunnel turning radius are also shown in Figure 2. The longitudinal length of the tunnel is defined as the length of the tunnel centerline. Thermocouples of K type are used for temperature measuring, which are placed 0.01 m below the tunnel ceiling. The thermocouple intervals in the longitudinal direction are set to be 0.2 m. To accurately measure the maximum temperature, a smaller longitudinal interval of 0.066 m and 0.033 m is adopted near the fire source.

As shown in Figure 2, the tunnel section between the entrance and the fire source is defined as being upstream of the fire source. The tunnel section between the exit and the fire source is defined as being downstream of the fire source. The artificial fire source system consists of a propane cylinder, a fuel pipeline, a mass flow controller, and a steel burner. The outlet size of the steel burner is 0.03 m × 0.03 m. During the experiment, the fire source is placed on the centerline of the tunnel at a distance of 1.4 m from the tunnel entrance. High-purity propane is adopted to produce artificial fire, whose supply rate is monitored by a digital mass flow controller.

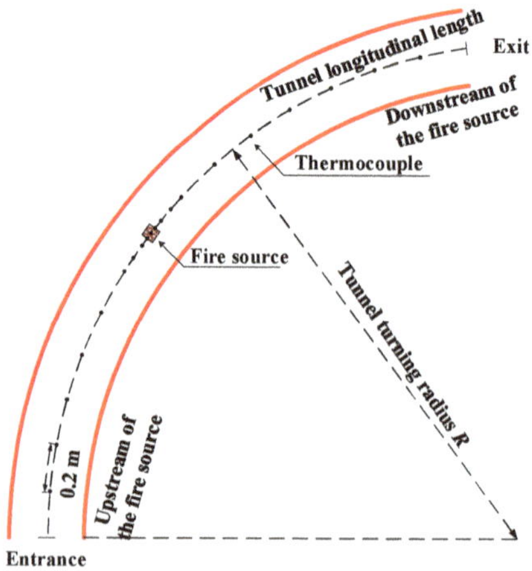

Figure 2. Thermocouple arrangement in a small-scale curved tunnel.

2.2. Experimental Scenarios

Table 1 summarizes the scenarios enacted in the small-scale experiments. A total of 28 scenarios were enacted, in which four different turning radiuses and seven different heat release rates are taken into account. The tunnel turning radius $R = \infty$ denotes a straight tunnel.

Table 1. The experimental scenarios conducted in the small-scale curved tunnel.

Scenarios No.	Fire Heat Release Rates (kW)	Tunnel Turning Radius R (m)
1–4	0.76	
5–8	1.01	
9–12	1.62	
13–16	2.03	3.5, 4.35, 8, ∞
17–20	2.43	
21–24	3.05	
25–28	4.06	

3. Results and Discussion

3.1. Variations in Maximum Temperature

Figure 3 depicts an instantaneous flame image observed in the experiment with different heat release rates. The phenomenon of flame plume impingement on the tunnel ceiling is observed even at smaller heat release rates. For instance, for $\dot{Q} = 0.76$ kW and 1.01 kW, the flame tip is close to the ceiling, and the thermocouple above the fire source is in direct contact with the flame.

Figure 4 plots the measured value of maximum temperature rise versus the turning radius, from which the influence of different turning radiuses can be recognized. The dotted lines represent the average temperature of different turning radiuses for the given heat release rate. It has been disclosed that the measured maximum temperature for different tunnel turning radiuses is nearly identical, which also indicates that the tunnel turning radius has little impact on the variation in maximum temperature.

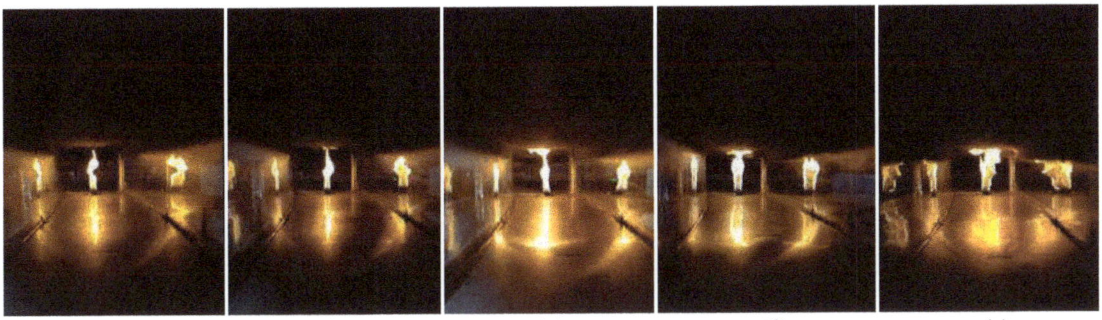

(**a**) (**b**) (**c**) (**d**) (**e**)

Figure 3. Instantaneous flame image with different HRRs: (**a**) $\dot{Q}$ = 0.76 kW; (**b**) $\dot{Q}$ = 1.01 kW; (**c**) $\dot{Q}$ = 1.62 kW; (**d**) $\dot{Q}$ = 2.43 kW; (**e**) $\dot{Q}$ = 4.06 kW.

The phenomenon of fire plume impingement in the tunnel can be clearly observed for all scenarios in the experiment. The flame height is almost equivalent to the source-ceiling height, i.e., the ceiling jet is driven by a strong plume. The temperature directly above the fire source mainly relates to the temperature of the impinged plume, which is one of the reasons why the tunnel turning radius has little effect.

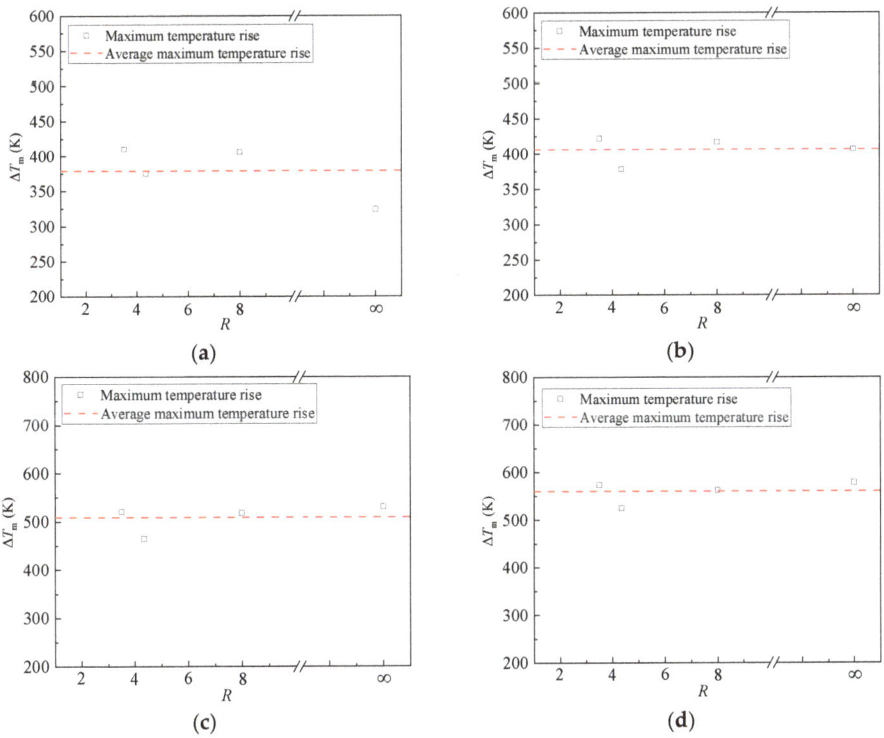

Figure 4. *Cont.*

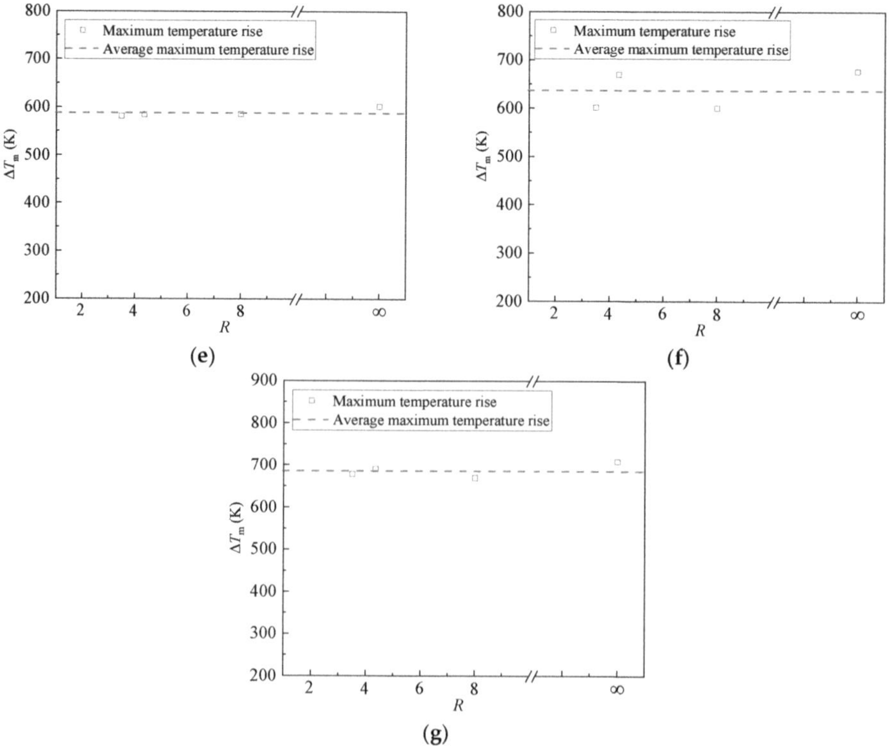

Figure 4. Measured value of maximum temperature rise (ΔT_m) against the tunnel turning radius: (**a**) $\dot{Q}$ = 0.76 kW; (**b**) $\dot{Q}$ = 1.01 kW; (**c**) $\dot{Q}$ = 1.62 kW; (**d**) $\dot{Q}$ = 2.03 kW; (**e**) $\dot{Q}$ = 2.43 kW; (**f**) $\dot{Q}$ = 3.05 kW; (**g**) $\dot{Q}$ = 4.06 kW.

The measured maximum temperature of all the experimental scenarios is compared with the predicted model proposed by Li et al. [9]. Figure 5 plots the measured value of maximum temperature as a function of $\dot{Q}^{2/3} H_{ef}^{5/3}$, along with Equation (5). It seems that the measured temperature in this work is significantly higher than the prediction based on Equation (5). This deviation is because Equation (5) is based on the theory of a "weak plume", whereas the ceiling jet condition in this work is more in accordance with the "strong plume driven" condition. As a result, the measured data cannot be accurately described by previously given correlation.

A predicted correlation of the temperature rise along the fire plume was proposed by McCaffrey and is given as [26]:

$$2g\frac{\Delta T}{T_\infty} = \left(\frac{\kappa}{C}\right)^2 \cdot \left(\frac{z}{\dot{Q}^{2/5}}\right)^{2\eta-1} \quad (7)$$

where z is the distance from the fire source in the vertical direction (m), ΔT is the temperature rise at the height of z (K), and κ, C, and η are constants whose values are shown in Table 2.

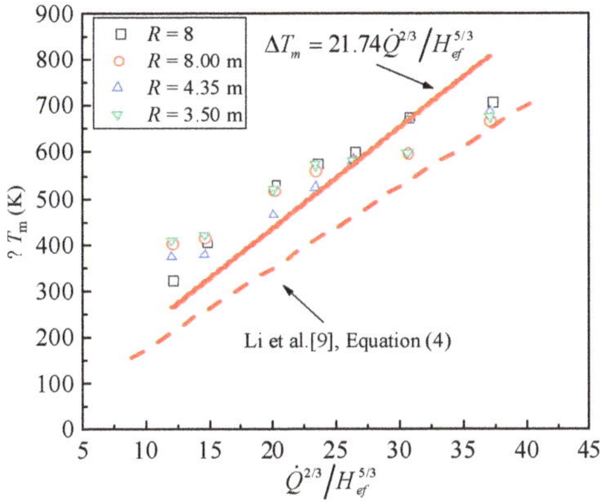

Figure 5. Prediction of maximum temperature rise (ΔT_m) based on Equation (5).

Table 2. Empirical constants in Equation (7).

Zone	κ	C	η
Continuous flame	6.9	0.9	1/2
Intermittent flame	1.9	0.9	0
Buoyant plume	1.1	0.9	−1/3

The fire plume temperature at the same height as the ceiling is given as:

$$\Delta T_m \propto \Delta T = \frac{T_\infty}{2g}\left(\frac{\kappa}{C}\right)^2 \cdot \left(\frac{z}{\dot{Q}^{2/5}}\right)^{2\eta-1}. \tag{8}$$

Equation (8) can be given in a dimensionless form as:

$$\frac{\Delta T_m}{T_\infty} \propto \left(\frac{H_{ef}/D_h}{\dot{Q}_s^{*2/5}}\right)^{\alpha}, \tag{9}$$

$$\dot{Q}_s^* = \frac{\dot{Q}}{\rho_\infty T_\infty c_p \sqrt{g} D_h^{5/2}}, \tag{10}$$

where: D_h is the hydraulic diameter of the burner outlet (m), α is constant, and $\dot{Q}_s^*$ is the heat release rate in dimensionless form using D_h as a characteristic length.

Figure 6 plots the measured maximum temperature rise versus the heat release rate in dimensionless form based on Equation (9), in which a good agreement between Equation (9) and experimental data is shown. The correlation in Figure 6 can be given as:

$$\frac{\Delta T_m}{T_\infty} = 4.79\left(\frac{H_{ef}/D_h}{\dot{Q}_s^{*2/5}}\right)^{-1} 1.8 \le \frac{H_{ef}/D_h}{\dot{Q}_s^{*2/5}} \le 3.7 \tag{11}$$

The tunnel ceiling dimensionless maximum temperature rise is related to −1 power of the distance between the fire surface and the ceiling in the vertical direction H_{ef}, indicating that the intermittent flame zone impinges on the ceiling, which is consistent with the phenomenon observed in our experiments.

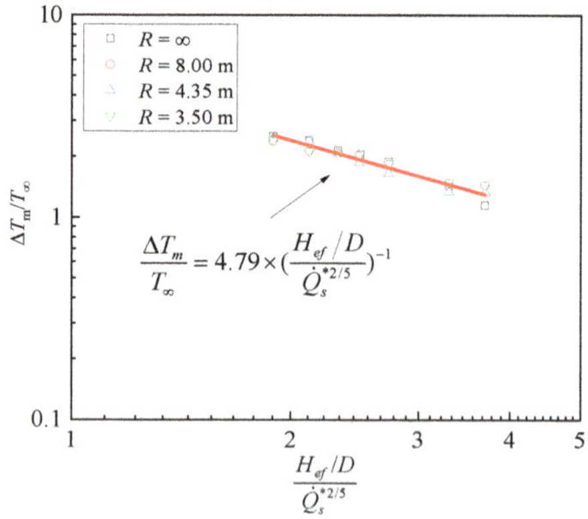

Figure 6. Predicted correlation of maximum temperature for the scenario of strong plume-driven ceiling jet.

3.2. The Longitudinal Attenuation of Temperature

The longitudinal attenuation of the temperature below ceilings is depicted in Figure 7 for a straight tunnel ($R = \infty$) with various heat release rates, where d denotes the longitudinal distance between the fire source and the temperature measuring point. For a given measuring position, the measured temperature rise (ΔT_d) increases with the heat release rate. As the distance from the fire source increases, the temperature decreases gradually. The temperature attenuation in both the upstream and downstream direction were found to be nearly identical. Therefore, only the temperature attenuation in the downstream direction is discussed.

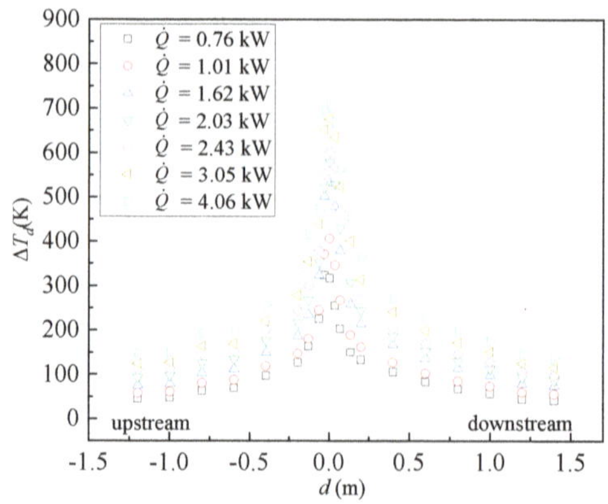

Figure 7. Longitudinal attenuation of the temperature of the longitudinal direction in a straight tunnel ($R = \infty$).

Figure 8 depicts the longitudinal attenuation of the temperature below ceilings with different turning radiuses. The temperature decay rate between a straight tunnel and a curved tunnel is similar for a given heat release rate, which indicates that the turning radius has little influence on smoke spread under natural ventilation.

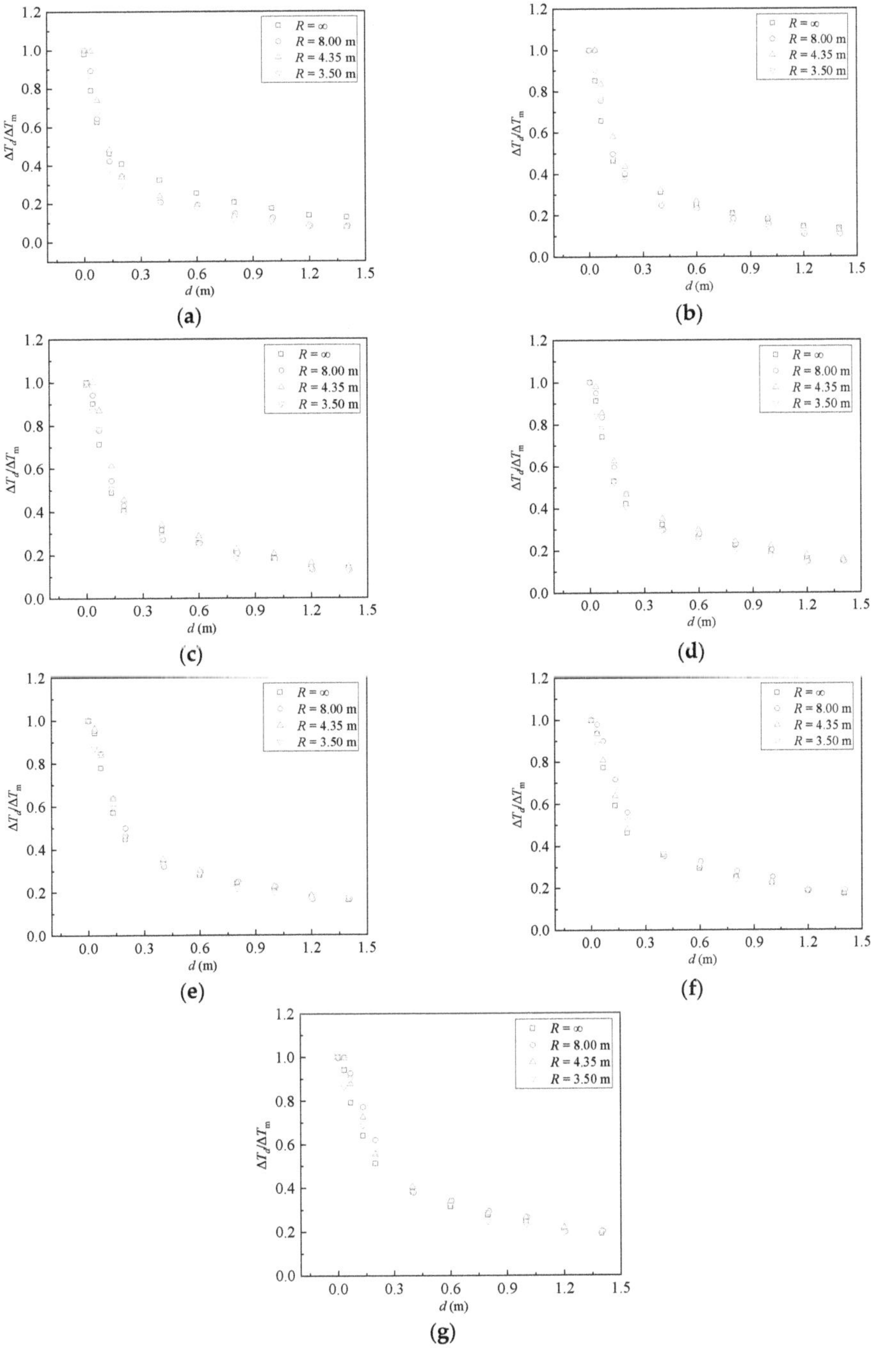

Figure 8. Longitudinal attenuation of temperature below the tunnel ceiling with different turning radiuses: (**a**) ΔT_m) against the tunnel turning radius: (**a**) $\dot{Q}$ = 0.76 kW; (**b**) $\dot{Q}$ = 1.01 kW; (**c**) $\dot{Q}$ = 1.62 kW; (**d**) $\dot{Q}$ = 2.03 kW; (**e**) $\dot{Q}$ = 2.43 kW; (**f**) $\dot{Q}$ = 3.05 kW; (**g**) $\dot{Q}$ = 4.06 kW.

Figure 9 shows the temperature attenuation of all the experimental scenarios based on Equation (6). The scattering of the experimental data can be clearly observed. It may be suspected that the measurement points from which the maximum temperature is measured cannot serve as the reference point (ΔT_0) for the condition of strong plume impingement since it is not within the region of "one-dimensional flow".

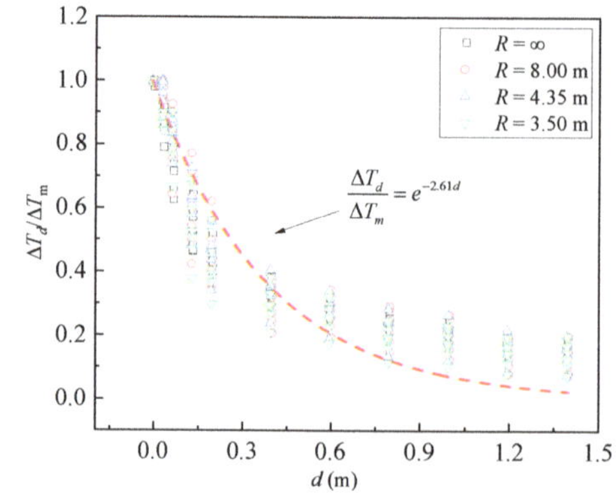

Figure 9. Plotting of temperature attenuation based on Equation (6).

To establish the relationship between the maximum temperature and longitudinal temperature attenuation, Heskestad and Hamada's [27] correlation is introduced, which was originally used for the temperature attenuation under an unconfined ceiling. It is given as:

$$\frac{\Delta T_r}{\Delta T_m} = 1.92\left(\frac{r}{b}\right)^{-1} - e^{[1.61(1-\frac{r}{b})]} \qquad 1 < \frac{r}{b} < 40 \tag{12}$$

$$b = 0.42\left[(\rho_\infty c_p)^{4/5} T_\infty^{3/5} g^{2/5}\right]^{-1/2} \frac{T_m^{1/2} \dot{Q}_c^{2/5}}{\Delta T_m^{3/5}}, \tag{13}$$

where r represents the distance between the fire source and the measuring point in the horizontal direction (m), b represents the characteristic length of the fire plume (m), and $\dot{Q}_c$ is the heat release rate of convection (kW).

Figure 10 plots the longitudinal attenuation of temperature in dimensionless form based on Equation (12). The longitudinal attenuation of temperature in the longitudinal direction of all scenarios collapsed well.

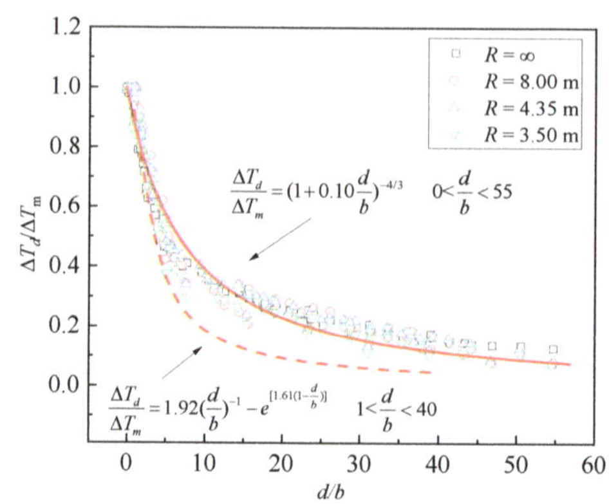

Figure 10. Prediction correlation of longitudinal attenuation of temperature below a ceiling for a scenario with a strong plume-driven ceiling jet.

Equation (12) has also been plotted in Figure 10, where a significant difference can be found between Equation (12) and the experimental data. The reason for such a discrepancy may be attributed to the fact that Equation (12) is proposed for temperature attenuation below an unconfined plate. For fire-induced smoke flow spreading in the tunnel, a region of one-dimensional spreading is formed where the mechanism of smoke spread is distinctly different from that under an unconfined ceiling. According to the fitting results of the experimental data, Equation (14) is more appropriate for predicting the temperature decay in a strong plume-driven tunnel fire. The improved correlation may be more applicable to a large vehicle fire in naturally ventilated tunnels, where the flames can easily impinge on the tunnel ceiling.

A layer of fresh air will be formed that flows to the fire source during tunnel fires under natural ventilation [28]. Studies concerning induced air spread are of interest and will be further studied.

$$\frac{\Delta T_d}{\Delta T_m} = (1 + 0.1\frac{d}{b})^{-4/3} 0 < \frac{r}{b} < 55. \tag{14}$$

4. Conclusions

The distribution characteristics of temperature below a curved tunnel ceiling induced by fire is studied through a small-scale tunnel model. The maximum temperature and longitudinal attenuation of temperature are analyzed. The following are some of the conclusions that may be drawn:

(1) The tunnel turning radius has limited influence on the maximum temperature and the longitudinal attenuation of the temperature below the ceiling in a strong plume impingement scenario. The temperature distribution profiles induced by a strong plume in a curved tunnel have similar tendencies to those observed in a straight tunnel (Figures 4 and 8).

(2) The maximum temperature below the ceiling under strong plume impingement cannot be well-predicted by previously proposed equations based on weak plume assumptions. This is because the thermocouple is in direct contact with the flame and the measured temperature is related to the axial temperature of the fire plume at the ceiling height (Figures 5 and 6).

(3) Improved correlation for predicting the temperature attenuation along the tunnel are proposed based on Heskestad and Hamada's model, which is capable of predicting the temperature decay along the tunnel under a strong plume-driven ceiling jet (Figure 10).

Author Contributions: H.T.: Writing—original draft and funding acquisition. Z.X.: project administration, resources, and supervision. D.Z.: data curation, investigation, validation, and visualization. All authors have read and agreed to the published version of the manuscript.

Funding: This work was supported by the National Key R&D Program of China (Grant No. 2017YFB1201204), the Hunan Provincial Innovation Foundation for Postgraduates (Grant No. CX20190131), and the Fundamental Research Funds for the Central Universities of Central South University (Grant No. 2022zzts0746).

Institutional Review Board Statement: Not applicable.

Informed Consent Statement: Not applicable.

Acknowledgments: We acknowledge the High Performance Computing Center of Central South University for its support.

Conflicts of Interest: The authors declare no conflict of interest.

References

1. Du, T.; Yang, D.; Peng, S.N.; Xiao, Y.M. A method for design of smoke control of urban traffic link tunnel (UTLT) using longitudinal ventilation. *Tunn. Undergr. Space Technol.* **2015**, *48*, 35–42. [CrossRef]
2. Ren, F.; Shi, C.L.; Li, J.; Che, H.L.; Xu, X. Numerical study on the flow characteristics and smoke temperature evolution under double fires condition with a metro train in tunnel. *Tunn. Undergr. Space Technol.* **2021**, *114*, 104553. [CrossRef]
3. Zhang, X.C.; Hu, L.H.; Zhu, W.; Zhang, X.L.; Yang, L.Z. Flame extension length and temperature profile in thermal impinging flow of buoyant round jet upon a horizontal plate. *Appl. Therm. Eng.* **2014**, *73*, 15–22. [CrossRef]

4. Du, T.; Yang, D.; Peng, S.N.; Liu, Y.L.; Xiao, Y.M. Performance evaluation of longitudinal and transverse ventilation for thermal and smoke control in a looped urban traffic link tunnel. *Appl. Therm. Eng.* **2016**, *96*, 490–500. [CrossRef]
5. Tang, F.; He, Q.; Chen, L.F.; Li, P.C. Experimental study on maximum smoke temperature beneath the ceiling induced by carriage fire in a tunnel with ceiling smoke extraction. *Sustain. Cities Soc.* **2019**, *44*, 40–45. [CrossRef]
6. Alpert, R.L. Calculation of response time of ceiling-mounted fire detectors. *Fire Technol.* **1972**, *8*, 181–195. [CrossRef]
7. Heskestad, G.; Delichatsios, M.A. The initial convective flow in fire. *Symp. Int. Combust.* **1979**, *17*, 1113–1123. [CrossRef]
8. Delichatsios, M.A. The flow of fire gases under a beamed ceiling. *Combust. Flame* **1981**, *43*, 1–10. [CrossRef]
9. Li, Y.Z.; Lei, B.; Ingason, H. The maximum temperature of buoyancy-driven smoke flow beneath the ceiling in tunnel fires. *Fire Saf. J.* **2011**, *46*, 204–210. [CrossRef]
10. Hu, L.H.; Huo, R.; Li, Y.Z.; Wang, H.B.; Chow, W.K. Full-scale burning tests on studying smoke temperature and velocity along a corridor. *Tunn. Undergr. Space Technol.* **2005**, *20*, 223–229. [CrossRef]
11. Li, L.M.; Cheng, X.D.; Wang, X.G.; Zhang, H.P. Temperature distribution of fire-induced flow along tunnels under natural ventilation. *J. Fire Sci.* **2012**, *30*, 121–136. [CrossRef]
12. Huang, Y.B.; Li, Y.F.; Dong, B.Y.; Li, J.M.; Liang, Q. Numerical investigation on the maximum ceiling temperature and longitudinal decay in a sealing tunnel fire. *Tunn. Undergr. Space Technol.* **2018**, *72*, 120–130. [CrossRef]
13. Zhang, X.L.; Lin, Y.J.; Shi, C.L.; Zhang, J.P. Numerical simulation on the maximum temperature and smoke back-layering length in a tilted tunnel under natural ventilation. *Tunn. Undergr. Space Technol.* **2021**, *107*, 103661. [CrossRef]
14. Zhang, T.H.; Wang, G.Y.; Hu, H.H.; Huang, Y.D.; Zhu, K.; Wu, K. Study on temperature decay characteristics of fire smoke backflow layer in tunnels with wide-shallow cross-section. *Tunn. Undergr. Space Technol.* **2021**, *112*, 103874. [CrossRef]
15. Tang, F.; Li, L.J.; Chen, W.K.; Tao, C.F.; Zhan, Z. Studies on ceiling maximum thermal smoke temperature and longitudinal decay in a tunnel fire with different transverse gas burner locations. *Appl. Therm. Eng.* **2017**, *110*, 1674–1681. [CrossRef]
16. Tang, F.; Chen, L.; Chen, Y.H.; Pang, H.P. Experimental study on the effect of ceiling mechanical smoke extraction system on transverse temperature decay induced by ceiling jet in the tunnel. *Int. J. Therm. Sci.* **2020**, *152*, 106294. [CrossRef]
17. Ingason, H.; Li, Y.Z. Model scale tunnel fire tests with longitudinal ventilation. *Fire Saf. J.* **2010**, *45*, 371–384. [CrossRef]
18. Chen, L.; Tang, F. Experimental study on the longitudinal temperature decay beneath ceiling in ventilated tunnel fire. *J. Therm. Anal. Calorim.* **2020**, *139*, 3179–3184. [CrossRef]
19. Tao, L.L.; Zeng, Y.H.; Li, J.; Yang, C.C.; Fang, Y.; Li, B.J. Study on the maximum temperature and temperature decay in single-side centralized smoke exhaust tunnel fires. *Int. J. Therm. Sci.* **2022**, *172*, 107277. [CrossRef]
20. Tang, F.; Cao, Z.L.; Chen, Q.; Meng, N.; Wang, Q.; Fan, C.G. Effect of blockage-heat source distance on maximum temperature of buoyancy-induced smoke flow beneath ceiling in a longitudinal ventilated tunnel. *Int. J. Heat Mass Transf.* **2017**, *109*, 683–688. [CrossRef]
21. Zhang, S.G.; Yang, H.; Yao, Y.Z.; Zhu, K.; Zhou, Y.; Cheng, X.D. Numerical Investigation of Back-Layering Length and Critical Velocity in Curved Subway Tunnels with Different Turning Radius. *Fire Technol.* **2017**, *53*, 1465–1793. [CrossRef]
22. Wang, F.; Wang, M.N.; Carvel, R.; Wang, Y. Numerical study on fire smoke movement and control in curved road tunnels. *Tunn. Undergr. Space Technol.* **2017**, *67*, 1–7. [CrossRef]
23. Xu, Z.S.; Zhou, D.M.; Tao, H.W.; Zhang, X.C.; Hu, W.B. Investigation of critical velocity in curved tunnel under the effects of different fire locations and turning radiuses. *Tunn. Undergr. Space Technol.* **2022**, *126*, 104553. [CrossRef]
24. Tang, F.; He, Q.; Sun, X.P.; Jiang, L.; Hu, P.; Hu, L.H. Experimental study of carriage fire in a tunnel: Evolution of flame geometry characteristics under relative strong crosswinds. *Proc. Combust. Inst.* **2021**, *38*, 4963–4970. [CrossRef]
25. Li, Y.S.; Zhang, X.L.; Sun, X.P.; Zhu, N. Maximum temperature of ceiling jet flow in longitudinal ventilated tunnel fires with various distances between fire source and cross-passage. *Tunn. Undergr. Space Technol.* **2021**, *113*, 103953. [CrossRef]
26. Hu, L.H.; Wang, Q.; Tang, F.; Delichatsios, M.A.; Zhang, X.C. Axial temperature profile in vertical buoyant turbulent jet fire in a reduced pressure atmosphere. *Fuel* **2013**, *106*, 779–786. [CrossRef]
27. Heskestad, G.; Hamada, T. Ceiling Jets of Strong Fire Plumes. *Fire Saf. J.* **1993**, *21*, 69–82. [CrossRef]
28. He, L.; Xu, Z.S.; Markert, F.; Zhao, J.M.; Xie, E.; Liu, Q.L.; Fan, C.G. Ceiling jet velocity during the whole process of fire development in a tunnel. *J. Wind Eng. Ind. Aerodyn.* **2021**, *212*, 104588. [CrossRef]

fire

Article

Numerical Simulation on Smoke Temperature Distribution in a Large Indoor Pedestrian Street Fire

Weidong Lin [1], Qiyu Liu [2], Meihong Zhang [3], Bihe Cai [4], Hui Wang [5], Jian Chen [6] and Yang Zhou [2,*]

[1] Fujian Provincial Institute of Architectural Design and Research Co., Ltd., Fuzhou 350001, China
[2] School of Civil Engineering, Central South University, Changsha 410075, China
[3] Xiamen Fire Rescue Detachment, Xiamen 361013, China
[4] YanGuo Technology Co., Ltd., Xiamen 361001, China
[5] Fujian Construction Engineering Group Co., Ltd., Fuzhou 350001, China
[6] China Academy of Building Research, Beijing 100013, China
* Correspondence: zyzhou@csu.edu.cn

Abstract: In order to study the characteristics of fire smoke spread and temperature distribution of a large indoor pedestrian street under different heat release rates and smoke exhaust modes, this paper focuses on the analysis of fire smoke spread, visibility, smoke temperature distribution and variation curves in an atrium. This paper uses a numerical simulation method to conduct research. PyroSim fire simulation software is used to calculate this project, which is based on a full-scale experimental design scheme. The numerical simulation results show that under the conditions of higher heat release rate, the smoke spread rate is greater than that under the conditions of lower heat release rate. Furthermore, the average temperature of smoke in the atrium is also greater, up to about 400 °C. The conditions of a higher heat release rate also show the characteristics of faster generation, faster spread and a larger volume of smoke. When the smoke exhaust system is turned on, the thickness of the smoke layer and the smoke temperature decrease. There then comes a situation where the stabile section of the fire ends in advance. The simulation results of vertical temperature distribution in an atrium can fit the modified McCaffrey plume model in any case. Under all cases, the smoke temperature reaches the maximum directly above the fire source. The horizontal dimensionless smoke temperature rises under the atrium roof, and decreases exponentially with the dimensionless distance from the fire source. The greater the heat release rate of fire source is, the smaller the attenuation coefficient is, with a more than 50% change. When the smoke exhaust system is turned on, the smoke flow accelerates and the smoke is cooled rapidly. Thus, the attenuation coefficient increases. Additionally, the effect of mechanical smoke exhaust is better than natural smoke exhaust, because mechanical smoke exhaust makes air flow and heat exchange more intense. The variation amplitudes of the attenuation coefficient under natural smoke exhaust and mechanical smoke exhaust are 13% and 22%, respectively.

Keywords: safety engineering; numerical simulation; large indoor pedestrian street fire; smoke spread; temperature distribution; smoke exhaust

Citation: Lin, W.; Liu, Q.; Zhang, M.; Cai, B.; Wang, H.; Chen, J.; Zhou, Y. Numerical Simulation on Smoke Temperature Distribution in a Large Indoor Pedestrian Street Fire. *Fire* **2023**, *6*, 115. https://doi.org/10.3390/fire6030115

Academic Editor: Dahai Qi

Received: 3 February 2023
Revised: 3 March 2023
Accepted: 7 March 2023
Published: 13 March 2023

1. Introduction

With the rapid development of the social economy and the advancement of modernization, urban economies have gradually developed in the direction of diversification to meet the growing needs of people for a better life. Thus, a large number of urban commercial complexes have emerged. These commercial complexes integrate entertainment, catering, fitness, shopping, offices and other functions, which greatly meet the needs of life, and bring much comfort and convenience. Indoor pedestrian streets are commonly found in various complexes and are one of their core architectural elements, found in various forms, and also in different lengths and heights. Due to the unique narrow and long building structure of the indoor pedestrian street, and the large number of combustibles in it coupled with the dense population and high mobility, once a fire occurs, the fire is very likely to spread to the atrium. It will spread further into the adjacent buildings through the pedestrian street walkway, causing irreparable losses. On 15 February 2004, a particularly serious fire accident occurred in Zhongbai Commercial Building in Jilin, Jilin Province. The fire was caused by the fire of a warehouse built nearby the commercial building which then spread to it. The evacuation exit was blocked, resulting in 54 deaths and more than 70 injuries. On 30 June 2012, a major fire accident occurred in the Ryde Commercial Building in Jixian, Tianjin, resulting in 10 deaths and 16 serious injuries, which was caused by electrical sparks due to overloaded air conditioning. In recent years, fire accidents in shopping malls and pedestrian streets are emerging one after another, and their serious fire risks have attracted great attention from society. Therefore, it is an obligation to research the laws of smoke spread and temperature distribution in atriums and indoor pedestrian streets, to conduct the decisive factor for personnel safety evacuation [1,2].

The spread of smoke in pedestrian streets and large atriums, the laws of temperature distribution and the effect of smoke exhaust systems have always been the focus of research of many scholars at home and abroad. Hu et al. [3] obtained the characteristics of smoke spread and the temperature distribution law of fire below the ceiling by conducting full-scale fire experiments in a long walkway. Huo [4] and others conducted fire experiments in a large-space fire experimental hall, and preliminarily studied the mechanical smoke extraction efficiency of large-space building fires. The full-scale experiment can realistically reproduce the fire scene and achieve a high degree of matching with real fire, which is very reliable. However, because of many large experimental scenes, the experiment is quite difficult, being both laborious and time-consuming. Thus, the limitations of full-scale experiments are reflected. In recent years, CFD numerical simulation technology has matured and improved, bringing not only good visualization effects, but also high fidelity, in addition to including many parameters that cannot be measured by experiments. The technology has gradually become the mainstream research method for studying large-space fires. Rho [5] and Mowrer et al. [6] used numerical simulation technology to study the smoke spread characteristics of a fire in a large-space atrium. Hadjisophocleous et al. [7] combined solid experiments and numerical simulation to analyze the smoke spread characteristics of atrium fires and the calculation method of smoke layer height. Long et al. [8] carried out full-scale fire experiments to investigate some key parameters, including the vertical and longitudinal temperature distribution, smoke layer height and smoke front arrival time under four different cases. In Jiao's work, a series of full-scale fire tests were performed to study the smoke spread characteristics and temperature distribution of indoor pedestrian street fires with different mechanical smoke exhaust modes [9]. Tian and Cai [10] used FDS to analyze and evaluate the fire hazard, effectiveness of smoke prevention measures and personnel safety of commercial complexes, and proposed performance-based design solutions. Zhao [11] studied the law of fire smoke spread and the control mode and effect of smoke in buildings through FDS numerical simulation technology, and obtained the influence of different smoke exhaust positions, fire source power, floor opening rate, smoke trapping wall height, etc., on smoke spread and the effects of smoke extraction. Jiang [12] used Fluent software to study the influence of the position of the air outlet on the smoke flow and smoke extraction effect of the atrium space of the mall, and proposed an optimization scheme.

It can be seen from the research status at home and abroad that the research on smoke control theory and smoke extraction mode has been relatively detailed. Although many studies have been carried out on large-space indoor pedestrian street fires, most of them focus on the influence of different smoke exhaust locations or smoke control modes on the flow and control of fire smoke. There are few studies on the temperature distribution characteristics of fire smoke in large-space indoor pedestrian streets. Therefore, this paper aims to study and discuss the characteristics of atrium smoke spread, vertical smoke temperature distribution and horizontal smoke temperature distribution under the ceiling under different ignition heat release rates, and establish a dimensionless theoretical model of smoke temperature rise.

2. Numerical Simulation Parameter Setting and Experimental Verification

2.1. Introduction of PyroSim

Fire numerical simulation research technology began in the 1960s. Over decades of continuous development and breakthrough, relatively mature research methods have been formed. Currently, fire numerical simulation research has been widely used in building fire research, performance building fire design, personnel safety escape and evacuation and other aspects. In this paper, PyroSim simulation software was selected to study the fire characteristics of indoor pedestrian streets for numerical simulation. This software is a fire dynamic simulation software based on the principle of large eddy simulation (LES), researched and developed by the National Institute of Standards and Technology (NIST). PyroSim is based on computational fluid dynamics and provides fire simulation parameters and fire model settings with a visual interface, which can accurately simulate and predict fire smoke spread, temperature, visibility, toxic and harmful gas concentration distribution and other parameter indicators [13,14].

2.2. Commercial Pedestrian Street Experiment

The experimental area was located in a pedestrian street and adjacent ring corridor in a building complex in Fuzhou, as shown in Figure 1. The building has 4 floors, each with a floor height of 4.5 m, and the pedestrian street is a narrow and straight, 154 m long and 17 m wide. Corridors on both sides are 4 m wide with a height of 2.8 m under the ceiling. The pedestrian street contains three atriums 1#, 2# and 3#. All three atriums are 19.8 m high. The experimental location is Atrium 2#, which is 32.4 m long and 8 m wide. The doors and windows on both sides of the ring corridor were kept closed during the experiment. The smoke exhaust system was opened according to different working conditions of the experiment, and the volume of the top smoke exhaust fan was 64,200 m^3/h.

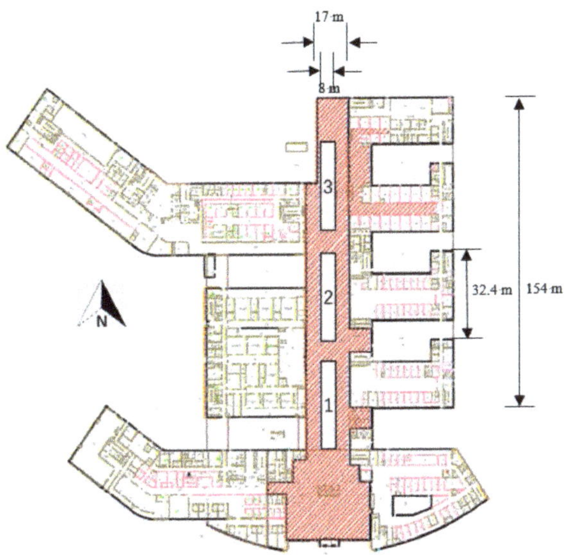

Figure 1. Plan of a building complex in Fuzhou.

The experimental working conditions include natural filling, natural smoke exhaust and atrium mechanical smoke exhaust. Specific working conditions are detailed in Table 1.

Table 1. Experiments' conditions.

Case	Heat Release Rates	Fire Location	Smoke Exhaust Mode
Case 1	1.5 MW	Atrium 2#	natural filling
Case 2	1.5 MW	Atrium 2#	natural smoke exhaust
Case 3	0.7 MW	Atrium 2#	natural filling
Case 4	0.34 MW	Atrium 2#	natural smoke exhaust
Case 5	0.34 MW	Atrium 2#	mechanical smoke exhaust

During the experiment, the fire source was arranged in the center of atrium 2#, as shown in Figure 2 in accordance with the requirements of the standard hot smoke test method [15] specified in the 'Field Verification Method for the Performance of the Anti-smoke System' (GA/T 999-2012). The designed heat release rates are 1.5 MW, 0.7 MW and 0.34 MW.

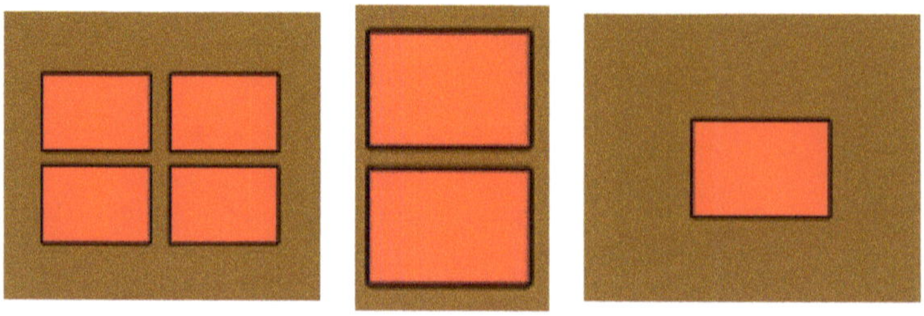

Figure 2. Schematic diagram of fire source layout of different heat release rates (1.5 MW, 0.7 MW, 0.34 MW, from left to right).

The basic components of the whole set of devices include the combustion plate placed in the water tray and the smoke generating device close to the combustion plate. The heat generated by the combustion of the fire source drives the tracer smoke movement to simulate the smoke spread behavior under the real fire. The combustion disc in the hot smoke test device was welded from 1.6 mm thick steel plate, and the internal dimensions were 841 mm × 595 mm × 130 mm. The plate area was 0.5 m^2, 95% ethanol was selected as fuel, 16 L was injected for each test. Cooling water was injected into the water-bearing plate to ensure the test safety, and the combustion plate will not float. Ammonium chloride smoke cake was selected as the smoke generating material for the test. The smoke generated by combustion was guided to the top of the fire source through an independent smoke generator. The tracer smoke was neutral, white, and basically free of residue.

Five thermocouples were arranged directly above the fire source. The bottom thermocouple was 0.9 m away from the fire source, and the next ones were arranged every next 0.35 m upward, numbering 1~5 in order. The top thermocouple was 2.3 m away from the fire. From the No. 5 thermocouple, they then diverged to the south, east, north and west directions, and along that axis, 2 thermocouples were arranged in each direction at intervals of 0.35 m, numbered 6~13 in turn. There were 13 thermocouples in total.

A thermocouple was arranged every 1 m at 2.3 m from the fire source. The highest point was 17.3 m, and they were numbered V1~V16 from top to bottom, to a total of 16 thermocouples. 7 thermocouples were arranged along the center line from west to east at the height of 9.3 m, 13.8 m and 18.3 m. The horizontal distance interval was 1 m. thermocouples were numbered Z3F1~Z3F7, Z4F1~Z4F7, ZTH1~ZTH7 from west to east in order, to a total of 21. A total of 37 thermocouples were used in this experiment. See Figure 3 for thermocouple layout.

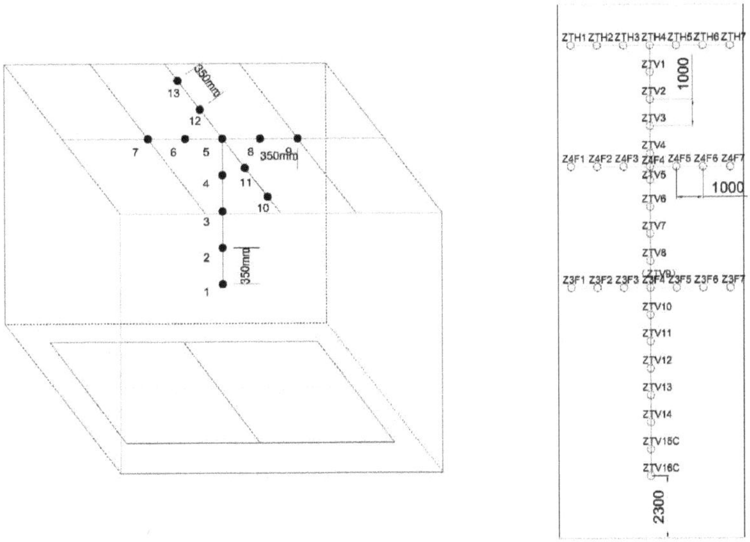

Figure 3. Thermocouple layout.

2.3. *Model Parameter Setting*

According to the building entity, a numerical model was established according to the size of 1:1. The wall material is made of concrete material and glass curtain wall, and all details of the model are consistent with the actual situation. The corresponding measurement points were set at each place of the thermocouples determined in the laboratory, and the temperature slices are placed in the X and Y directions at the center point of the fire source. The design heat release rates are 1.5 MW, 0.7 MW and 0.34 MW. The fire model is a t^2 fast fire. The simulated initial temperature was 20 °C. The simulation time was consistent with the experimental time, and all doors and windows were kept closed.

The smoke exhaust vents were set in the smoke storage bin on the roof of the building and consisted of 52 vents. A total of 20 natural smoke exhaust vents with lengths of 1.5 m and heights of 0.4 m were set on the east and west sides. Furthermore, 6 mechanical smoke exhaust vents with lengths of 1.2 m and heights of 0.4 m were set on the north and south sides.

During the series of tests, the start-up time of the mechanical smoke exhaust system was 30 s after ignition, and the ambient temperature was always maintained at about 20 °C. The next group of tests were conducted after the environmental conditions recovered to the initial state.

The schematic diagram of numerical model of this complex is shown in Figure 4.

2.4. *Grid Independence Analysis*

When FDS is used for fire simulation, the mesh size has a greater impact on the experimental results. The smaller the mesh size, the smaller the numerical fluctuation of the simulation results, the more accurate the experimental results, and the longer the simulation time. Thus, it is very important to select the appropriate mesh size for fire simulation. The dimensionless expression $D^* = \delta_x$ is given in the FDS Operation Manual [16], and δ_x is the nominal size of the grid cell. Its definition formula is as follows:

$$D^* = \left[\frac{Q}{\rho_0 c_p T_0 \sqrt{g}}\right]^{2/5} \tag{1}$$

where D^* is the characteristic diameter of fire, m; Q is the heat release rate, kW; g is the acceleration of gravity, m/s^2; ρ_0 is the ambient air density, 1.29 kg/m^2; c_p is the specific heat capacity at constant pressure, 1.005 kJ/(kg·K); T_0 is the ambient air temperature, 293 K. Taking the heat release rate of 1.5 MW as an example, the characteristic size of the fire is D^* = 1.09 m. It is generally believed that when the ratio of characteristic diameter to grid is 4~16, the simulation results are more accurate, that is, the grid size is 0.27~0.06 m. Due to the large volume of the physical model, the grid size of the atrium is assumed to be 0.3 m, which basically meets the operation conditions.

Figure 4. Schematic diagram of numerical model of a building complex.

2.5. Experimental Verification of Numerical Simulation Results

As is can be seen from Figure 5, under the same working conditions, the experimental results are in good agreement with the numerical simulation results. However, the experimental temperature data will drop sharply at a certain moment Through observing the experimental video, we know that the phenomenon is caused by the phenomenon of flame fusion and separation during the experiment, so that the temperature data of the measurement point in the center will fluctuate greatly, which is a normal phenomenon. From the overall temperature change trend, this effect can be ignored, and the data error is within an acceptable range. Therefore, we believe that the calculation results of the numerical simulations in this paper are trustworthy.

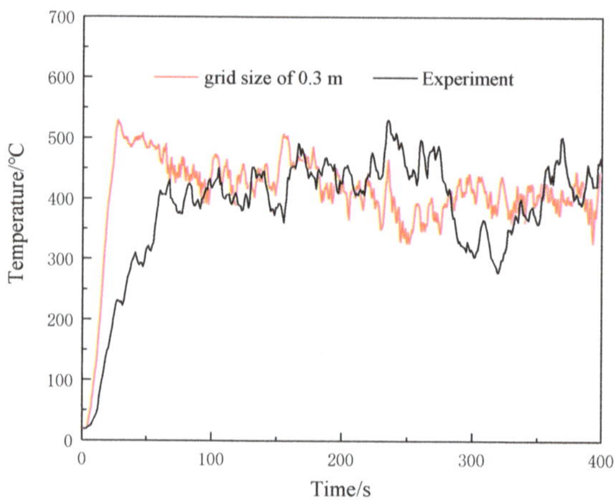

Figure 5. Comparison of temperature change between experiment and numerical simulation of measuring point 1 in Case 1.

Analyzing the two sets of data of numerical simulations and experiments for variance, the sum of three squared errors was calculated to construct the test statistic: SSA $= \sum_{i=1}^{r}\sum_{j=1}^{n}\left(\overline{x}_i - \overline{\overline{x}}\right)^2$, SSE $= \sum_{i=1}^{r}\sum_{j=1}^{n}\left(\overline{x}_{ij} - \overline{x}_i\right)^2$. The total deviation squared sum SST = SSA + SSE, and the corresponding degrees of freedom are r − 1, n − r, n − 1, thus determining the mean square MSA, MSE and MST between groups. The required test statistics F can be obtained by the ratio of MSA and MSE, F = MSA/MSE. Based on significance level α, compare F and $F_\alpha(r-1, n-r)$ [17]. The calculation results of analysis of variance are shown in Table 2.

Table 2. Calculation results of analysis of variance.

Error Source	Sum of Squares	Freedom	Mean Square	F	$F_\alpha(r-1, n-r)$
Intergroup differences	2.08291×10^6	1	2.08291×10^6	113.26094	3.85
Intragroup differences	2.21971×10^7	1027	18390.333		
Sum	2.428×10^7	1028			

By calculation, it can be seen that at the significance level of 0.05, F is significantly greater than F_α, showing a significant difference between the two sets of data.

3. Analysis and Discussion of Numerical Simulation Results

3.1. Analysis of Smoke Spread in Atrium Fire

A large amount of smoke and heat generated during a fire will form a hot smoke stream. The flow direction of the smoke is often the direction of the fire spread, and the flow speed of the smoke is often the fire spreading speed [18]. The smoke will gradually collect over the atrium, and the smoke layer will continue to settle, which will continuously reduce visibility, affect the visual range of evacuees and then affect the evacuation speed.

According to the principle of fire dynamics, the development of fire goes through three stages named accelerated combustion, stable combustion and the extinguishing stage. The spread of smoke in each stage is also different. In the stage of accelerated flame combustion, fire smoke is generated and continues to spread upwards and accumulates, reaching the ceiling and continuing to spread around. The fire has developed into a stable combustion stage, and a large amount of smoke generated before has accumulated in the ceiling and formed a stable smoke layer, which continues to settle. Until the extinguishing phase, smoke is continuously generated and fills the entire atrium. Figure 6 shows the schematic diagram of smoke spread under working Cases 1~5.

It is not difficult to see from Figure 6 that the smoke rises first under the action of thermal buoyancy, and after reaching the ceiling, it begins to spread horizontally until it fills the ceiling and spreads into the ring corridor. Due to the restrictions of the walls on the east and west sides of the atrium, the smoke begins to fill downward, and then the smoke continues to be generated. At 300 s, it has basically filled the atrium, and the smoke layer in the atrium begins to settle slowly, with a clear boundary with the cold air layer.

Comparing Cases 1, 3 and 4, we find that different heat release rates will make a significant difference in the speed of smoke spread; under the condition of a large heat release rate, the smoke spreading rate is greater than that of the small heat release rate, and the thickness of the stable smoke layer is also significantly thicker. It is not difficult to understand that the greater the heat release rate is, the more obvious the thermal buoyancy effect is, so the stronger the winding effect is. Therefore, it can reach the roof earlier, gather, and begin to settle. When the natural smoke exhaust (Cases 1, 2) and mechanical smoke exhaust (Cases 4, 5) are turned on, the concentration and thickness of the smoke layer will be significantly reduced. This is because the opening of the smoke exhaust vent makes the smoke continuously pump outside, reducing the concentration of the smoke layer and making the thickness of the smoke layer decrease. From this point of view, the smoke exhaust effect of the large atrium is good, and it is very necessary to set up a reasonable smoke exhaust system.

3.2. Vertical Smoke Temperature Distribution in Atrium

The fire produces a large amount of smoke, and releases huge heat to produce a high temperature. A high temperature environment and toxic smoke will bring difficulties to the safe evacuation of trapped people. Long-term exposure to smoke will seriously damage people's physical functions, and they may lose the ability to escape. The study of the temperature distribution law of smoke in an atrium can not only provide a reference for building fire design, but also provide guidance for personnel evacuation during a fire. Figures 7–11 are the variation charts of vertical smoke temperature in the atrium under five different working conditions, and six measurement points at different positions are selected as analysis objects in the vertical direction: V1 (17.3 m), V4 (14.3 m), V7 (11.3 m), V10 (8.3 m), V13 (5.3 m), V16 (2.3 m).

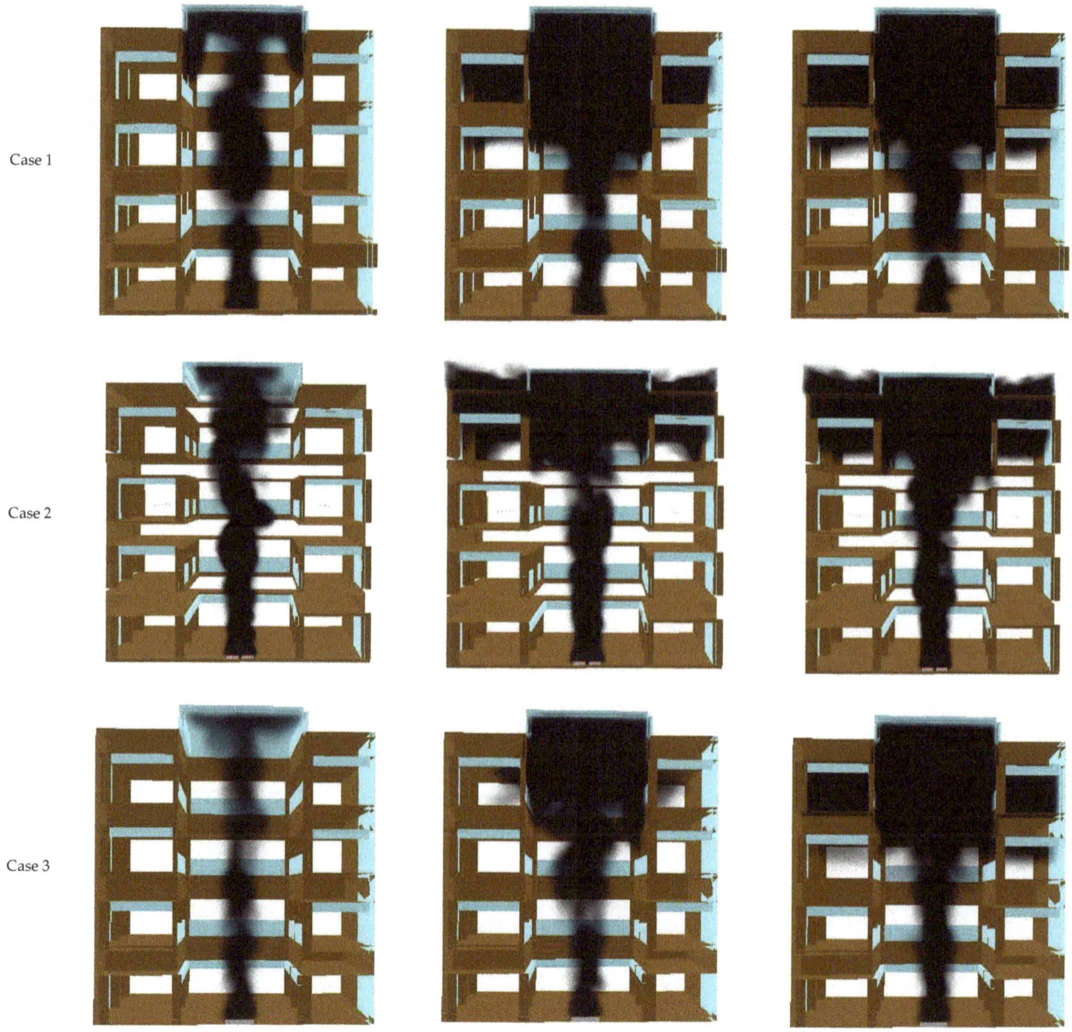

Figure 6. *Cont.*

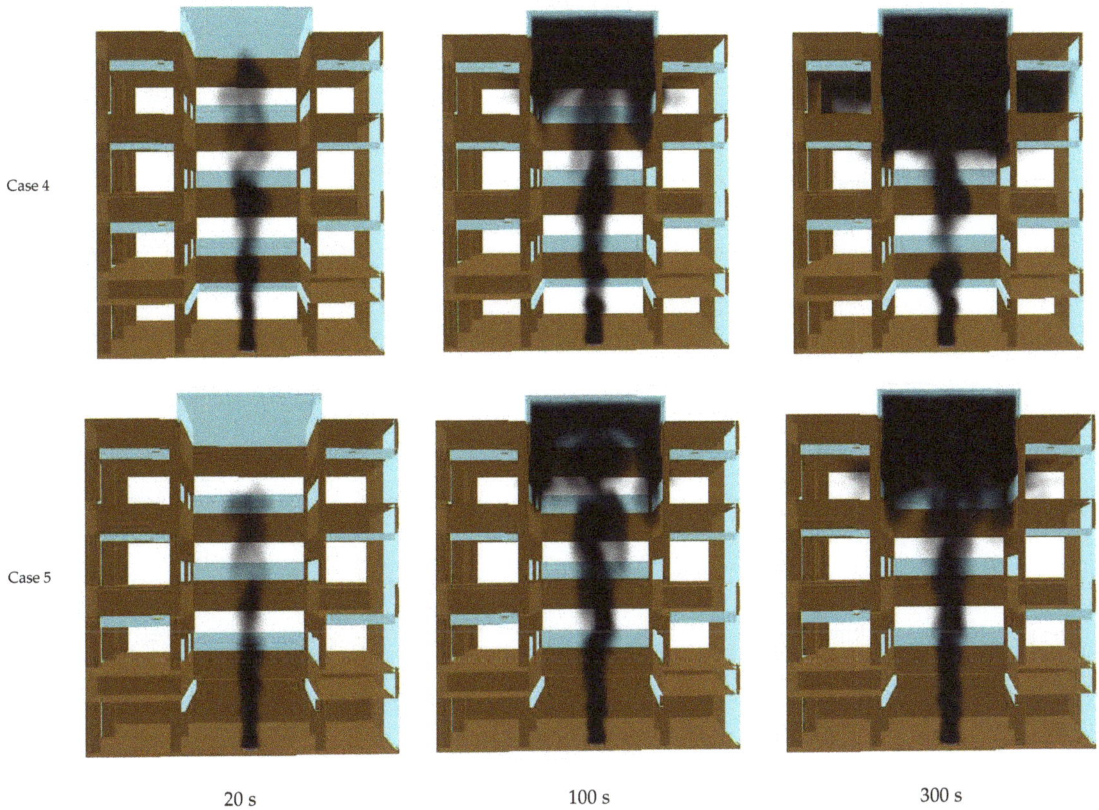

Figure 6. Comparison of smoke layer thickness at different times in Cases 1~5.

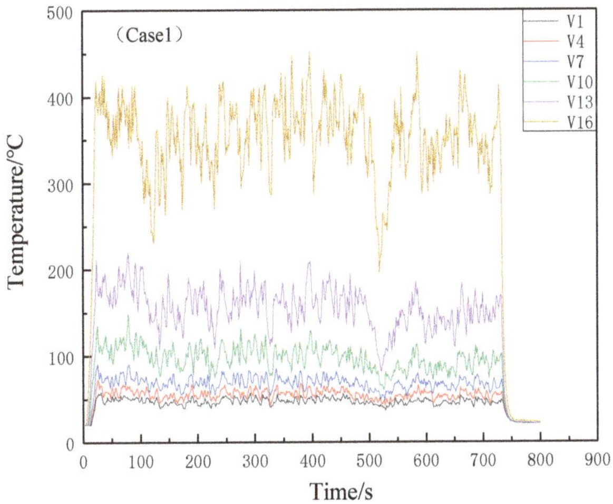

Figure 7. Temperature variation of measuring points at different heights above the atrium in Case 1.

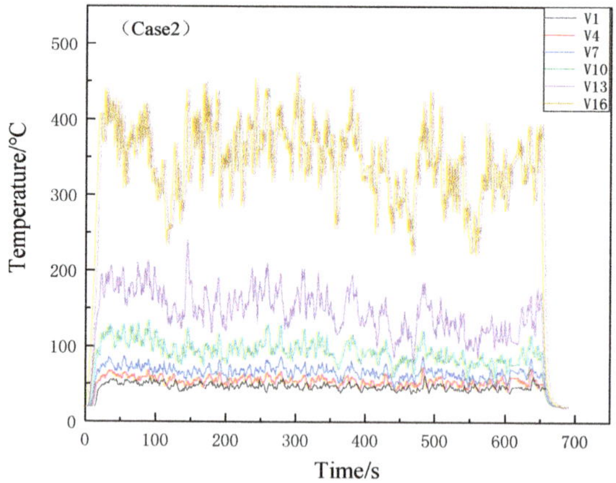

Figure 8. Temperature variation of measuring points at different heights above the atrium in Case 2.

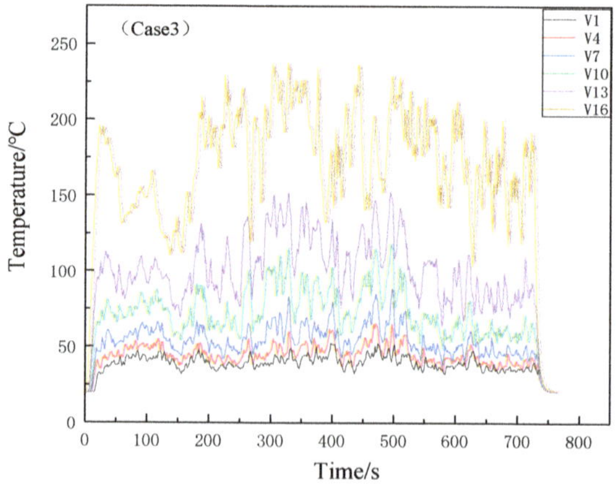

Figure 9. Temperature variation of measuring points at different heights above the atrium in Case 3.

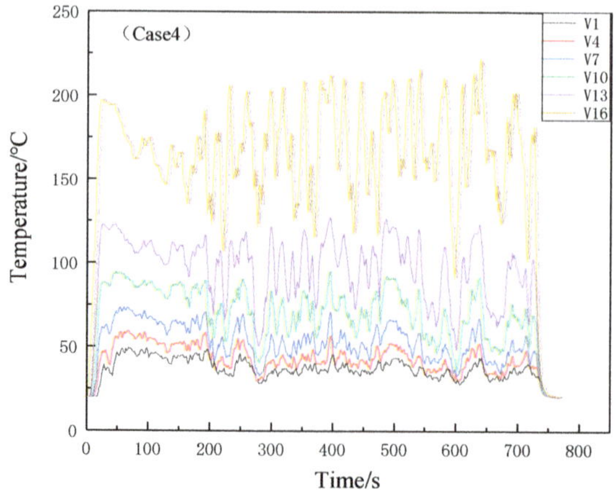

Figure 10. Temperature variation of measuring points at different heights above the atrium in Case 4.

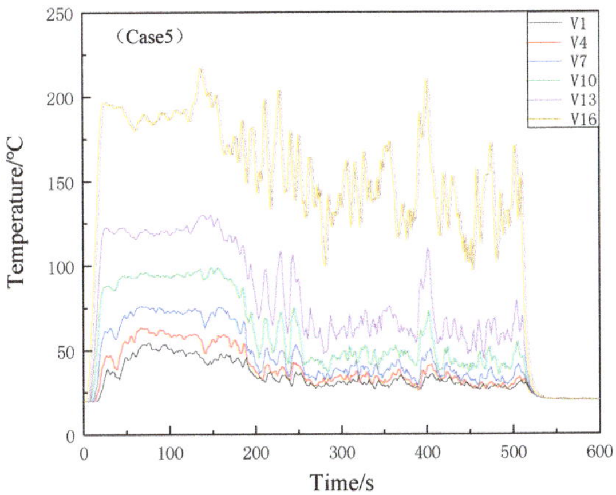

Figure 11. Temperature variation of measuring points at different heights above the atrium in Case 5.

It can be seen that the change trend of smoke temperature rise under different working conditions is basically similar. Regardless of the heat release rate and whether the smoke exhaust system is turned on, the rise of smoke temperature has successively experienced the three stages of rise, fluctuation within a certain range and decrease, which is consistent with the three stages of occurrence and development of fire explained by combustion theory. In the stable combustion stage of the ignition source, the smoke temperature will fluctuate within a certain range, and we regard it as the stable stage of the smoke temperature rise (Figures 7–11). Under different working conditions, the stabilization stage of smoke temperature rise is different, which is mainly reflected in the stable value of smoke temperature rise and the time when the stable section disappears.

When a fire starts, the heat released by the fire is limited, and the rise of smoke temperature is relatively slow. Then, the fire becomes violent, and the temperature gradient increases rapidly and it quickly reaches the maximum temperature. Comparing Figures 7, 9 and 10, it can be seen that under the same natural filling conditions, the temperature stability values of the measurement point closest to the fire source under different heat release rate in Case 1 (1.5 MW), Case 3 (0.7 MW) and Case 4 (0.34 MW) decrease sequentially. This indicates that the greater the heat release rate is, the greater the average temperature of the smoke in the atrium is. The greater the distance between the smoke and the fire source above the atrium, the lower the temperature of the smoke is; that is, the thermocouple closest to the fire source has the highest temperature and the thermocouple farthest away has the lowest temperature. As the smoke moves far away in the upper layers, heat is gradually dissipated. However, the time it takes for the smoke temperature stabilization to disappear is not much different. When the natural smoke exhaust mode (Cases 1 and 2) is turned on, it mainly has a greater impact on the smoke temperature on the upper floor of the atrium, and has little impact on the smoke temperature near the fire source. This is because the natural smoke exhaust strategy only opens the top glass window of the atrium to accelerate the flow of smoke near it, so that the heat exchange is accelerated, while the position near the fire source far from the glass window receives little impact. After turning on the mechanical smoke exhaust system (Cases 4 and 5), the temperature decreases, which also causes the temperature stability section to disappear earlier.

Figure 12 shows the change of the average temperature of the vertical smoke stabilization section of the atrium with the height of the fire source center under the conditions of Cases 1~5. It can be clearly known that the uniform temperature of the smoke stabilization section is directly related to the heat release rate of the fire. In addition, the use of natural smoke exhaust (Cases 1 and 2) or mechanical smoke exhaust (Cases 4 and 5) also has an impact on the uniform temperature of the smoke stabilization section. Turning on the smoke exhaust system will accelerate the heat exchange between the smoke and outside, so that the smoke cooling temperature is reduced. However, from Figure 12, the magnitude of this temperature reduction is not obvious, indicating that a single smoke exhaust mode has a poor effect on the flow control of smoke.

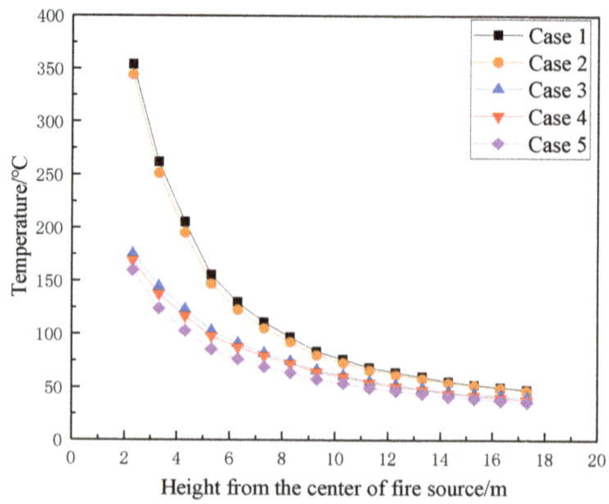

Figure 12. Variation of vertical smoke average temperature in atrium with height from fire source center in Cases 1~5.

In order to describe the distribution of vertical smoke temperature in the atrium more intuitively and accurately, we introduced the McCaffrey plume model [19]. The McCaffrey plume model is a semi-empirical formula, which is fitted by a large number of experimental results. It is applicable to the calculation of plume mass flow under the condition of both small area fires and large area fires, maintaining certain universality. It is widely used to study the flame plume in building fires [20–23]. The expression is as follows:

$$\Delta T = \left(\frac{\kappa}{0.9 \cdot \sqrt{2g}}\right)^2 \left(\frac{z}{\dot{Q}^{2/5}}\right)^{2\eta-1} \cdot T_\infty \tag{2}$$

where ΔT represents the difference between the temperature at appointed altitude and the ambient temperature, K; g is the acceleration of gravity, 9.81 m/s^2; $\dot{Q}$ is the heat release rate of the fire, kW; z is the height, m; T_∞ is the ambient temperature, 293 K. The three zones of the axisymmetric buoyant plume are shown in Figure 13. The values of κ and η of the McCaffrey plume model are shown in Table 3.

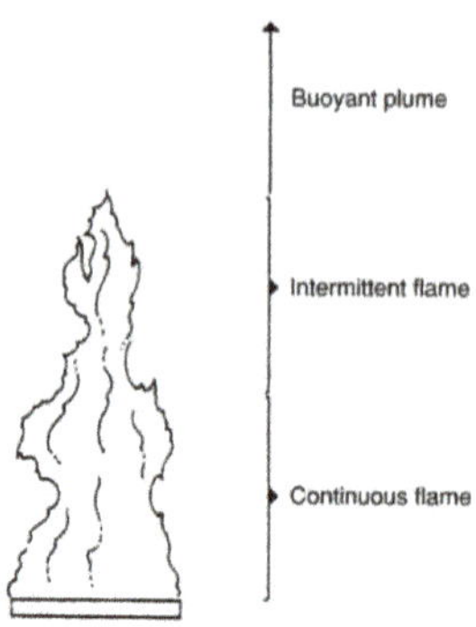

Figure 13. The three zones of the axisymmetric buoyant plume.

Table 3. Values of κ and η.

Zone	$z/\dot{Q}^{2/5}$ [m/kW$^{2/5}$]	κ	η
Continuous flame zone	<0.08	6.8 [m$^{1/2}$/s]	1/2
Intermittent flame zone	0.08–0.2	1.9 [m/kW$^{1/5}$ s]	0
Plume zone	>0.2	1.1 [m$^{4/3}$/kW$^{1/3}$ s]	−1/3

However, considering the factors of multi-fire flame fusion, air flow and smoke exhaust system in this working condition, the McCaffrey plume model above needs to be corrected. We still use $z/\dot{Q}^{2/5}$ as the independent variable to draw the scatter diagram of smoke temperature rise changing with the vertical height and the heat release rate of the fire source, and then carry out subsection fitting correction, as shown in Figure 14.

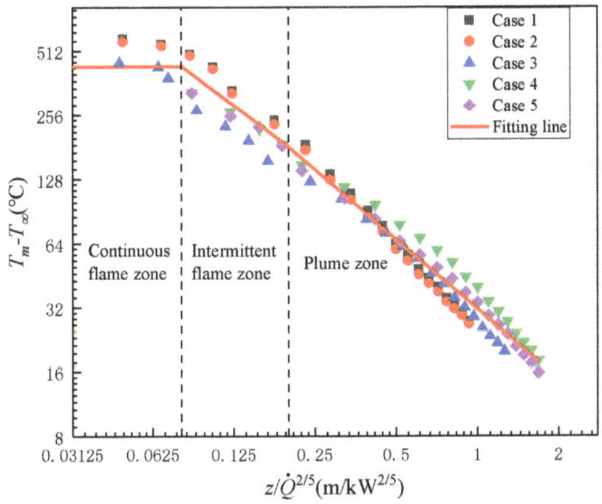

Figure 14. Variation of smoke temperature rise with $z/\dot{Q}^{2/5}$.

The modified κ and η values are shown in Table 4.

Table 4. Modified values of κ and η.

Zone	$z/\dot{Q}^{2/5}$ [m/kW$^{2/5}$]	κ	η
Continuous flame zone	<0.08	4.84 [m$^{1/2}$/s]	0.5
Intermittent flame zone	0.08–0.2	1.46 [m/kW$^{1/5}$ s]	0.03
Plume zone	>0.2	1.31 [m$^{4/3}$/kW$^{1/3}$ s]	−0.04

3.3. Horizontal Smoke Temperature Distribution under the Ceiling of the Atrium

The temperature distribution below the atrium ceiling is similar to ceiling jet. The smoke rises from the atrium fire source to the ceiling, and then spreads around, filling the entire atrium. Under all cases, the smoke temperature reached a maximum value directly above the fire source, and decreased on the east and west sides as it moved away from the center point, showing an exponential attenuation mode. The temperature attenuation rate is not only directly related to the heat release rate, but also has a certain relationship with the smoke exhaust system. In order to analyze this attenuation relationship more intuitively and qualitatively, we make the average temperature of the smoke under the atrium ceiling dimensionless according to the following equation [24],

$$\Delta T/\Delta T_0 = e^{-\alpha x/H} \quad (3)$$

where ΔT represents the temperature of the measurement point in the 'x' direction, °C; ΔT_0 represents the temperature at the center position directly above the fire source, °C; α is the attenuation coefficient of dimensionless smoke temperature rise with dimensionless distance from fire source; x is the distance from the center point, m; H is the clear height of atrium, 19.8 m.

Figure 15 shows the variation of the horizontal dimensionless smoke temperature rise $\Delta T/\Delta T_0$ with the dimensionless distance x/H from the fire source at the height of 18.3 m below the atrium ceiling under various working conditions. The correlation coefficient of the fitting curve reaches more than 0.95 under all cases, indicating that the consistency between the data and the model was good. Table 5 shows the attenuation coefficient of dimensionless smoke temperature rise with dimensionless distance from the fire source.

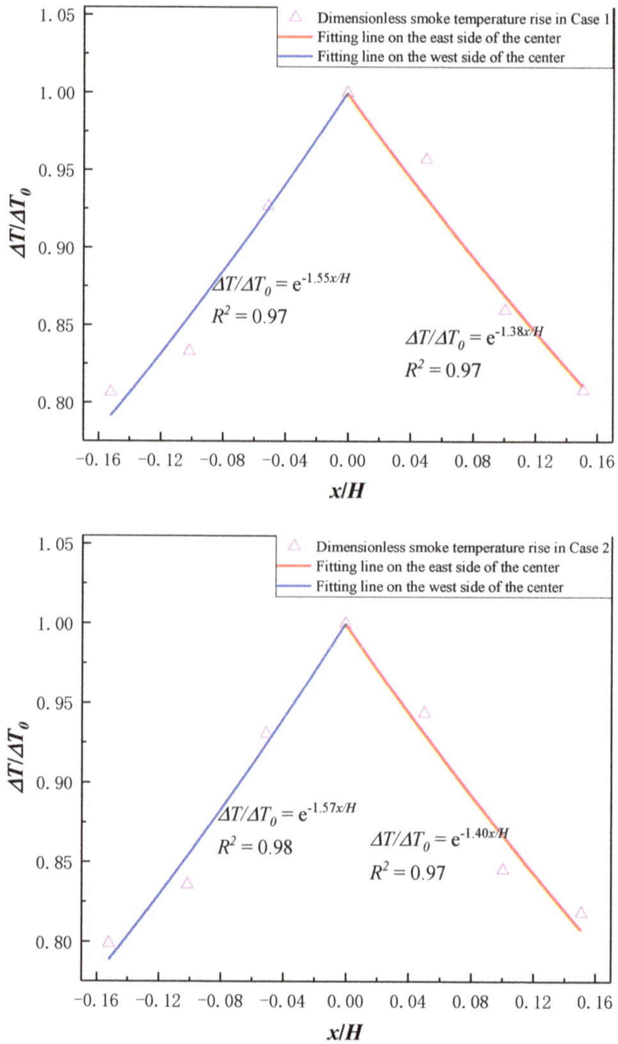

Figure 15. *Cont.*

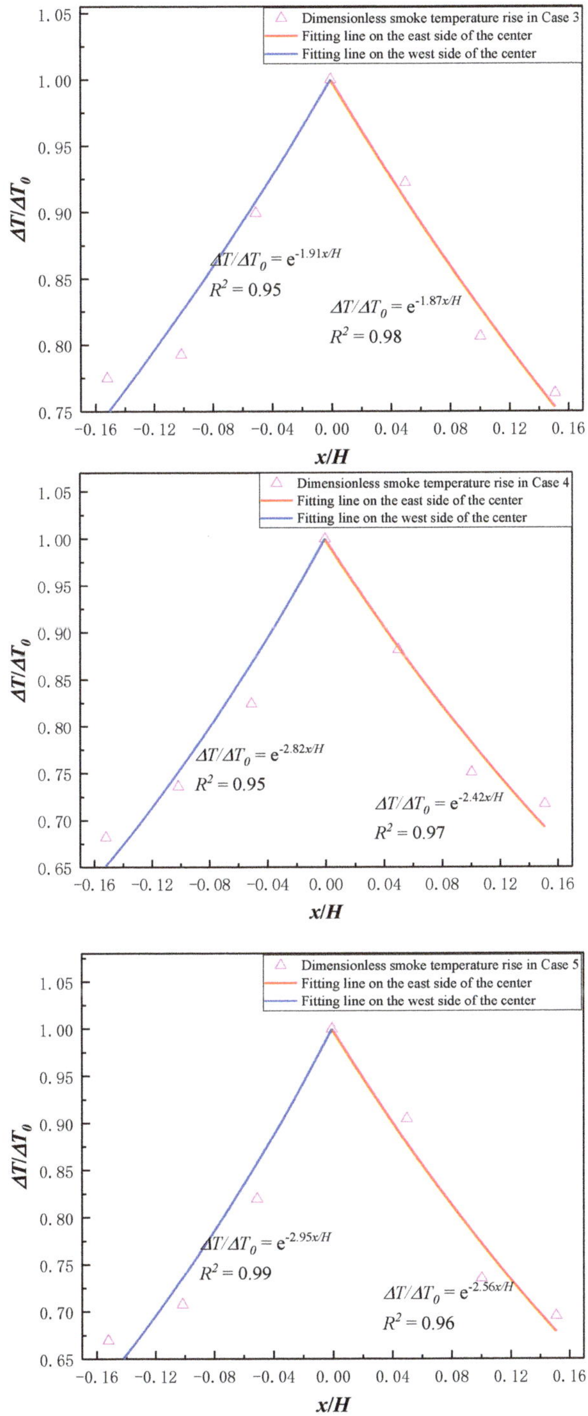

Figure 15. Variation of dimensionless smoke temperature rise $\Delta T/\Delta T_0$ under atrium ceiling with dimensionless distance from fire source *x/H* in Cases 1~5.

Table 5. Attenuation coefficients of dimensionless smoke temperature rise with dimensionless distance from fire source.

	Case 1	Case 2	Case 3	Case 4	Case 5
α (East)	1.38	1.40	1.87	2.42	2.56
α (West)	1.55	1.57	1.91	2.82	2.95

It can be seen from Figure 15 that the variation of lateral dimensionless smoke temperature rise under the atrium ceiling with the distance between dimensionless and fire source has a greater relationship with the heat release rate. The larger the heat release rate is, the higher the smoke temperature in the atrium is, and the greater the mass flow rate is, also the faster the smoke spreads, resulting in a smaller attenuation coefficient. The smoke temperature shows a trend of attenuation from the central position to both sides, and the left and right sides are basically symmetrically distributed, but there are still slight differences. Possibly because the east side of the building has a protruding space, as well as the obstruction of the smoke barrier, the east side of the smoke accumulates in a small area, and the heat is not dispersed, resulting in a smaller attenuation rate than the west side. Analyzing the influence of different experimental scenarios, it is found that the dimensionless temperature rise attenuation rate is also related to the use of the smoke exhaust system. Specifically, after opening the natural smoke exhaust vents (Cases 1 and 2) and turning on the mechanical smoke exhaust (Cases 4 and 5), the smoke flow is accelerated, and the heat exchange with the external environment is promoted, bringing the cooling effect to the smoke. Therefore, it will lead to the increase of the attenuation coefficient, which means that the attenuation rate also increases. Additionally, the effect of mechanical smoke exhaust is better than that of natural smoke exhaust.

4. Conclusions

In this paper, the smoke spread law and temperature distribution characteristics of large indoor pedestrian street fires are investigated using a combination of numerical simulation and full-scale experiments. The main conclusions are summarized as follows:

(1) After a series of comparative studies, it is found that the conclusions obtained from the numerical simulation study and the full-scale experimental study are in good agreement. The smoke spread rate increases with the increase of the heat release rate of the fire source, and the thickness of the stable smoke layer increases as well. When the smoke exhaust system is turned on, the smoke volume decreases rapidly, the smoke layer thickness decreases and the visibility increases.

(2) The higher the heat release rate of the fire source, the higher the average temperature of the smoke in the atrium. There exists an obvious stabilization phase of the smoke temperature. When the smoke exhaust system is turned on, the smoke temperature decreases and the stabilization phase of the smoke temperature is shortened. In the actual scenario, the effect of multi-system cooperative smoke exhaust is better than the smoke exhaust effect of single smoke exhaust system. The smoke temperature distribution obtained from the numerical simulation agrees well with the modified McCaffrey plume model.

(3) The horizontal dimensionless smoke temperature rise below the atrium ceiling changes exponentially with the dimensionless distance from the fire source. The greater the heat release rate, the greater the smoke mass flow rate and smoke spread rate, and the smaller the attenuation coefficient. The attenuation coefficient increases when the smoke exhaust system is turned on. Furthermore, this indicates that the effect of mechanical smoke exhaust is better than that of natural smoke exhaust. Among the factors related to the attenuation coefficient, the effect of the heat release rate of the fire is stronger than that of smoke exhaust. In practical applications, using low calorific value materials, reducing the stacking of combustibles and adjusting the

exhaust mode and volume all contribute to the increase of the attenuation coefficient to decrease the atrium temperature as soon as possible.

Author Contributions: Conceptualization, Q.L. and Y.Z.; methodology, Y.Z.; software, Q.L. and W.L.; validation, W.L., B.C. and Y.Z.; formal analysis, Q.L.; investigation, Q.L.; resources, W.L.; data curation, B.C.; writing—original draft preparation, Q.L.; writing—review and editing, Q.L. and Y.Z.; visualization, M.Z., H.W. and J.C.; supervision, W.L. and J.C.; project administration, M.Z.; funding acquisition, B.C. and H.W. All authors have read and agreed to the published version of the manuscript.

Funding: This work was supported by the Regional Development Science and Technology Program of Fujian Province (No. 2019Y3002) and Natural Science Foundation of Hunan Province (No. 2021JJ30860).

Institutional Review Board Statement: Not applicable.

Informed Consent Statement: Not applicable.

Data Availability Statement: Not applicable.

Conflicts of Interest: The authors declare no conflict of interest.

References

1. Xu, X.; Wang, Z.; Liu, X.; Ji, C.; Yu, N.; Zhu, H.; Li, J.; Wang, P. Study on Fire Smoke Control in Super-high Building Atrium. *Procedia Eng.* **2018**, *211*, 844–852. [CrossRef]
2. Wang, R.; Lan, X.; Xu, L. Smoke spread process under different heights based on numerical simulation. *Case Stud. Therm. Eng.* **2020**, *21*, 100710. [CrossRef]
3. Hu, L.H.; Huo, R.; Li, Y.Z.; Wang, H.B.; Chow, W.K. Full-scale burning tests on studying smoke temperature and velocity along a corridor. *Tunn. Undergr. Space Technol. Inc. Trenchless Technol. Res.* **2004**, *20*, 223–229. [CrossRef]
4. Huo, R.; Li, Y.; Yu, M.; You, F.; Zhou, Y. Preliminary study on mechanical smoke exhaust in large-space building fire. *Fire Sci. Technol.* **2001**, *5*, 440–452.
5. Rho, J.; Ryou, H. A numerical study of atrium fires using deterministic models. *Fire Saf. J.* **1999**, *33*, 213–229. [CrossRef]
6. Mowrer, F.W. Enclosure smoke filling revisited. *Fire Saf. J.* **1999**, *33*, 93–114. [CrossRef]
7. Rafinazari, A.; Hadjisophocleous, G.V. An investigation of the effect of make-up air velocity on smoke layer height with asymmetric openings and rotational air flow in atrium fires. *J. Build. Eng.* **2020**, *27*, 100933. [CrossRef]
8. Long, Z.; Liu, C.; Yang, Y.; Qiu, P.; Tian, X.; Zhong, M. Full-scale experimental study on fire-induced smoke movement and control in an underground double-island subway station. *Tunn. Undergr. Space Technol.* **2020**, *103*, 103508. [CrossRef]
9. Jiao, A.; Lin, W.; Cai, B.; Wang, H.; Chen, J.; Zhang, M.; Xiao, J.; Liu, Q.; Wang, F.; Fan, C. Full-scale experimental study on thermal smoke movement characteristics in an indoor pedestrian street. *Case Stud. Therm. Eng.* **2022**, *34*, 102029. [CrossRef]
10. Tian, H.; Cai, W. Design of fire smoke control in commercial indoor pedestrian street. *Fire Sci. Technol.* **2013**, *32*, 1349–1351.
11. Zhao, Z. *Research on Smoke Flow and Control of Fire in Indoor Pedestrian Street*; Central South University: Changsha, China, 2014.
12. Jiang, C. *Numerical Simulation Study on Fire Smoke Control in Atrium Space of a Shopping Mall*; Anhui Jianzhu University: Hefei, China, 2018.
13. McGrattan, K.; McDermott, R.; Weinschenk, C.; Forney, G. *Fire Dynamics Simulator Users Guide*, 6th ed.; Nist Special Publication: Washington, DC, USA, 2013.
14. Lv, S. *Fire and Escape Simulation Chinese Tutorial and Engineering Application*; Beijing Chemical Industry Press: Beijing, China, 2014.
15. *GA/T 999-2012*; Field Verification Method for Anti-Smoke System Performance Hot Smoke Test Method. Ministry of Public Security of the People's Republic of China. China Standards Press: Beijing, China.
16. *NIST Special Publication 1019, April 2015. Fire Dynamics Simulation (Version 6.2.0)—User's Guide*; National Institute of Standards and Technology: Gaithersburg, MD, USA, 2015.
17. Dai, J. Test and Correction of Heteroscedasticity in One-Way ANOVA. *Stat. Decis.* **2017**, *8*, 23–26.
18. Zhong, M.; Li, P.; Liu, T. Experimental study on fire smoke movement in a multi-floor and multi-room building. *Sci. China Ser. E Eng. Mater. Sci.* **2005**, *48*, 292–304. [CrossRef]
19. Karlsson, B.; Quintiere, J.G. *Enclosure Fire Dynamics*; CRC Press: Boca Raton, FL, USA, 1999.
20. Walter, W.J.; Glenn, P.F. *NISTTN1431, A Technical Reference for CFAST: An Engineering Tool for Estimating Fire and Smoke Transport*; National Institute of Standards and Technology, Building and Fire Research Laboratory: Gaithersburg, MD, USA, 2000; pp. 17–77.
21. Cheng, Y.; Chen, L.; Zhang, M. The Analysis and Estimation of the Plume Models in the Development Fires. *Fire Saf. Sci.* **2002**, *11*, 132–136.
22. Han, R.; Zhu, G.; Zhang, G.; Zhang, Y. Optimization Design of Smoke Extraction in the High-rise Building Wit Atrium. *Ind. Saf. Environ. Prot.* **2013**, *39*, 48–51.

23. Tao, Y.; Lu, K.; Chen, X.; Mao, S.; Ding, Y.; Zhao, Y. Experimental investigation on the temperature profile of large scale RP-5 aviation kerosene pool fire in an open space. *Fuel* **2020**, *264*, 116852. [CrossRef]
24. Gong, L.; Jiang, L.; Li, S.; Shen, N.; Zhang, Y.; Sun, J. Theoretical and experimental study on longitudinal smoke temperature distribution in tunnel fires. *Int. J. Therm. Sci.* **2016**, *102*, 319–328. [CrossRef]

fire

Article

Effect of Different Smoke Vent Layouts on Smoke and Temperature Distribution in Single-Side Multi-Point Exhaust Tunnel Fires: A Case Study

Liangliang Tao [1,2] and Yanhua Zeng [1,2,*]

[1] Key Laboratory of Transportation Tunnel Engineering, Ministry of Education, Southwest Jiaotong University, Chengdu 610031, China; taolliang@my.swjtu.edu.cn
[2] School of Civil Engineering, Southwest Jiaotong University, Chengdu 610031, China
* Correspondence: zengyhua68@swjtu.edu.cn

Abstract: In this paper, a numerical model verified by a 1:10 small-scale model test was used to study the effect of different smoke vent layouts on fire characteristics and smoke exhaust efficiency. The results show that the total smoke spread length is shortest when four smoke vents are opened near the fire source. If there are more than four smoke vents, some of them will only inhale fresh air rather than smoke. More seriously, some smoke vents will promote the spread of toxic smoke farther. Under different smoke vent layout schemes, the maximum temperature shows the same change trend with the increase in smoke exhaust volume (first increasing and then decreasing). When there are four smoke vents, the temperature field is in a good range compared with other schemes. If four smoke vents are opened, the total smoke exhaust efficiency is highest, and exhaust rate has little influence on total exhaust efficiency. Total smoke exhaust efficiency of the tunnel is more than 93.7% under different exhaust volumes, and the maximum difference of total smoke exhaust efficiency is less than 1.5% under different exhaust volume of Case "4". The exhaust volume has little influence on temperature decay beneath the ceiling, and a temperature attenuation model of a point exhaust tunnel with four smoke vents was proposed. For the single-side point exhaust tunnels, the number of smoke vents near the exhaust fan side shall not be more than that on the other side. Four smoke vents shall be opened in case of fire and the exhaust volume is 220 m^3/s with HRR of 30 MW.

Keywords: tunnel fire; smoke control; maximum temperature; temperature decay; exhaust efficiency

Citation: Tao, L.; Zeng, Y. Effect of Different Smoke Vent Layouts on Smoke and Temperature Distribution in Single-Side Multi-Point Exhaust Tunnel Fires: A Case Study. *Fire* **2022**, 5, 28. https://doi.org/10.3390/fire5010028

Academic Editors: Chuangang Fan and Dahai Qi

Received: 21 January 2022
Accepted: 16 February 2022
Published: 18 February 2022

1. Introduction

The danger of fire in a long and narrow space refers, mainly, to the damage to the building structure caused by high-temperature and toxic smoke [1–3]. Controlling the smoke in a certain area and discharging it out of the buildings in time will help to reduce the fire hazard [4–10]. Therefore, tunnel ventilation and smoke control are the primary issues in tunnel fire safety research. In recent years, ceiling exhaust ventilation has been widely used in tunnel fire. Ventilation data for underwater tunnels constructed by shield tunnelling machine in southern China were collected through field research, as shown in Table 1. Research on the application of point exhaust ventilation in tunnel fire smoke control has been reported extensively. Zhao et al. [11] analyzed the critical smoke exhaust rate and temperature distribution in two-point exhaust tunnels through a 1:20 model test combined with theoretical analysis and obtained the corresponding prediction model. Tang et al. [12] studied the effect of the longitudinal velocity on maximum temperature and found out a calculation model of maximum temperature in a smoke exhaust tunnel. They also explored the influence of smoke exhaust on the transverse temperature profile beneath ceiling, and established a unified temperature calculation model [13]. Tang and other colleagues [14] also studied the effects of longitudinal wind speed and exhaust rate on tunnel fires and proposed a calculation model of smoke distribution considering exhaust volume.

Table 1. Ventilation data for underwater tunnels.

City	Tunnel	Length (km)	Number of Smoke Vents Opened	Distance between Smoke Vent (m)
Wuhan	Donghu Tunnel	7.035	4	60
Wuhan	Sanyang Road Tunnel	4.32	6	60
Wuhan	Qingdao Road Yangtze River Tunnel	3.44	4	60
Hangzhou	Qianjiang Tunnel	4.45	6	60
Yangzhou	Shouxihu Tunnel	4.40	6	60
Shanghai	Shanghai Yangtze River Tunnel	8.90	3	60

Some scholars have also studied lateral ceiling exhaust tunnel fires. Zhu et al. [15] studied the changes of critical wind speed and maximum temperature when the smoke vent is located on the side wall and put forward some calculation models of maximum temperature and critical wind speed that can guide the tunnel fire ventilation design based on the test results. Wang et al. [16] studied the effect of exhaust volume and smoke vent area on smoke back-layering length through a model test and obtained an empirical formula for calculating the length of smoke back-layering considering in a point exhaust tunnel. He et al. and Jiang et al. investigated the characteristics of smoke entrainment at the vent and found that the HRR and smoke exhaust rate have the greatest influence on smoke entrainment coefficient [17,18]. Tao et al. [19,20] studied the smoke control and temperature distribution of a two-point exhaust tunnel and obtained some conclusions and temperature calculation models that could improve tunnel ventilation design.

The factors, such as smoke and temperature distribution, related to evacuating one- or two-point exhaust tunnels have been widely reported [21–23]. A tunnel fire will quickly produce a large amount of toxic smoke. It is difficult to discharge the smoke out of the tunnel by opening only one or two smoke vents. Therefore, the design of tunnel fire ventilation allows for multiple smoke vents to be opened at the same time. More importantly, the traditional smoke exhaust tunnel will discharge the smoke from both sides of the tunnel, which will cause waste for some short tunnels. A separate fan room needs to be built for tunnel ventilation, and the cost is often very expensive. If fan rooms are designed only on one side and the exhaust duct outlet of the tunnel on the other side is closed, it will reduce not only the cost of construction but also the number of axial-flow fans. It will also greatly reduce the maintenance cost of ventilation facilities during operation. This kind of smoke extraction scheme was adopted in the Mawan Tunnel in Shenzhen, China. Unfortunately, the fire characteristics in single-side point exhaust tunnels have not been reported.

To fill this gap, this paper verified the reliability of the numerical model based on the tunnel model. The effects of different smoke vent opening modes and smoke exhaust rate on the smoke distribution, maximum temperature, temperature attenuation, and smoke exhaust efficiency in the tunnel were studied based on the numerical results.

2. Numerical Modeling

2.1. Model Tunnel

The applicability of FDS (6.7) in tunnel fire calculation has been widely verified [24–27]. In this paper, the full-scale model tunnel is 600 m long, 13 m wide, and 6.5 m high. The exhaust duct is 1.7 m high and 13 m wide. The area of the exhaust vent is 14.93 m^2 and the distance between smoke vents is 60 m. The smoke exhaust position consists of two smoke vents with a size of 1 × 3.5 m, and there are six groups of smoke vents, as shown in Figure 1. The two smoke vents in each smoke exhaust position are always in the same state (open or closed). The vault and side walls of the tunnel are made of concrete. During the whole simulation process, the environmental temperature is 20 °C and the mesh boundary condition is "OPEN" [28]. The mass flow of gas and CO mass fraction at the smoke vent should also be measured. In order to monitor the temperature distribution of the tunnel vault, the thermocouple is arranged at 0.20 m below the tunnel ceiling and at intervals of 0.25 m.

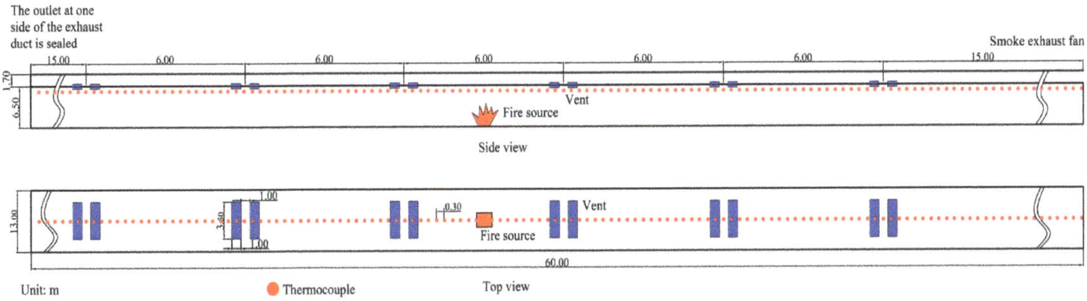

Figure 1. Schematic diagram of model tunnel.

Note that the HRR is closely related to the combustibles. The HRR caused by a car is 5–7 MW and that of medium-sized truck or bus is 20–30 MW. The HRR in a tunnel fire may be higher, such as when an oil tank truck is involved [21]. However, tunnel ventilation design must take into account the heat release rate according to the traffic flow and vehicle type after tunnel operation. Therefore, only one heat release rate is generally considered in tunnel fire ventilation design, and it is often much greater than the estimated value. In addition, China's national regulations (Guidelines for Design of Ventilation of Highway Tunnel, JTG/TD70/2-02-2014) clearly stipulate that the heat release rate of most highway tunnels in ventilation design should be 20 or 30 MW. Therefore, the power of the fire source is 30 MW, and the exhaust volume is 200–280 m^3/s in this paper. The fire source (gasoline) is located at the center line of the tunnel floor, and the power of gasoline is controlled by defining the mass loss rate (MLR) in FDS. The mass loss rate of gasoline is 0.055 $kg/(m^2 \cdot s)$, and the HRR can reach 30 MW when the area of fire source is 12.6 m^2 [29]. The soot and CO yield are set to 0.1 and 0.05, respectively [30]. A detailed calculation scheme is shown in Table 2.

Table 2. A summary of the simulation scheme.

Test No.	Smoke Vent Number	HRR (MW)	Smoke Exhaust Rate (m^3/s)
1–5	3A	30	200, 220, 240, 260, 280
6–10	3B	30	200, 220, 240, 260, 280
11–15	4	30	200, 220, 240, 260, 280
16–20	5A	30	200, 220, 240, 260, 280
21–25	5B	30	200, 220, 240, 260, 280
26–30	6	30	200, 220, 240, 260, 280

Note that a calculation model of temperature decay was also studied in this paper. If only one fire source heat release rate is considered, the temperature calculation model is not rigorous. Therefore, we studied the temperature attenuation beneath the tunnel ceiling with HRR of 10, 20, and 30 MW.

2.2. Mesh Size

Generally, mesh size is the most important aspect for numerical simulation because it determines the reliability of the numerical results. When the grid is less than $0.1D^*$, the numerical results are acceptable to guarantee the reliable operation of FDS [31]. The characteristic diameter D^* can be calculated by:

$$D^* = \left(\frac{Q}{\rho_a c_p T_a g^{1/2}} \right)^{2/5} \quad (1)$$

where T_a is the ambient air-temperature (K), ρ_a is the ambient air density (kg/m^3), c_p is the specific heat capacity of air at constant pressure ($kJ/kg \cdot K$), g is the gravity acceleration (m^2/s), and Q is the heat release rate of fire source (kW). D^* is calculated to be 3.74 m when the HRR of fire is 30 MW, thus $0.1D^*$ is approximately 0.374 m. In the previous reports on the use of FDS to study tunnel fires, the selection of mesh size was described in detail, and the mesh size was verified, as shown in Table 3. At present, when studying the smoke movement and temperature profile based on FDS, the commonly used grid sizes are 0.1667 and 0.20 m [30,32–36]. Since the tunnel length in this paper is 600 m, we set the mesh size as 0.20 m. Note that the HRR is the main factor affecting the characteristic diameter of the fire source. Since the research scenario in this paper is the same as the existing tunnel fire research based on FDS, we do not repeat the grid sensitivity analysis.

Table 3. Details of previous model tunnel grids.

References	Dimension (m × m) Wide × High	HRR (MW)	Mesh Sizes (m)
Yao [30]	10 × 5	5–100	0.200
Ji [33]	10 × 5	5–15	0.200
Liang [36]	10 × 7	30	0.20
Ji [37]	10 × 5	3–15	0.167
Guo [38]	10 × 5	3–10	0.167

2.3. Experimental Verification

A 1:10 model test with the same tunnel section and length was conducted to verify the reliability of the numerical model. The model tunnel is made of aluminum sheet and fireproof glass, and each smoke vent has a push–pull steel plate to control the opening state of the vent. The outlet on one side of the exhaust duct is connected with the smoke exhaust fan, and the outlet of the exhaust duct on the other side is sealed, as shown in Figure 2. The jet fan in Figure 2 was not installed during the test conditions in this paper. The fire source of the model test is liquefied petroleum gas (LPG), and the HRR of the fire source is controlled by a rotameter [39], as shown in Figure 2. Details of the small-scale model tunnel can be found in the author's previous research [20,40]. The HRR and smoke exhaust rate are converted by Froude similarity criterion.

Figure 2. Experimental device diagram.

In this paper, the numerical model will be verified from the temperature profile and smoke spread. The temperature profile in the tunnel is obtained under the same conditions. The temperature distribution of the numerical results is approximately the same as the model results, as shown in Figure 3a. The temperature difference between the numerical and model results is very small near the tunnel ceiling and floor but relatively large in the middle of the tunnel. The difference between the results of the model test and the numerical simulation is within 10.8 °C under the tunnel ceiling. Under the same condition, the smoke spread length in the numerical calculation is like that in the model test, and the maximum difference is 5.4 m. Maximum errors of temperature and back-layering length are 10.3% and 5.1%, respectively, which can be ignored for tunnel engineering. Therefore, the accuracy of the mesh size of the numerical model can be guaranteed.

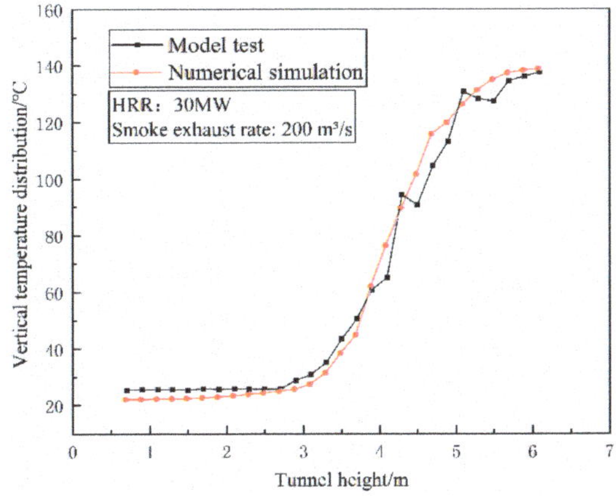

(a)

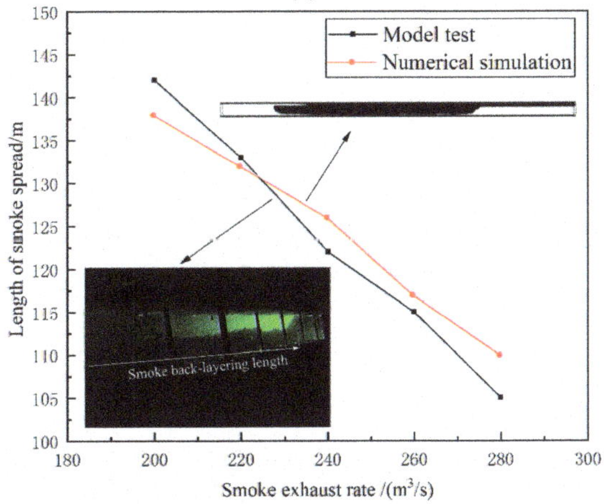

(b)

Figure 3. Comparison of experimental and numerical results at 30 MW of Case "4". (**a**) Temperature distribution. (**b**) Smoke spread length.

3. Results and Discussion

3.1. Smoke Spread

In a tunnel fire, the opening state of the smoke vent is 3, 4, 5, and 6. When opening three smoke vents, there are two situations (Figure 4): (1) Case "3A": two smoke vents are opened near the axial flow fan (upstream) and one is opened near the closed end (downstream); (2) one smoke vent is opened near the axial flow fan and two are opened near the closed end (Case 3B). There are also two situations when opening five smoke vents: (1) three smoke vents are opened near the axial flow fan (upstream) and two are opened near the closed end (downstream) (Case "5A"); (2) two smoke vents are opened near the axial flow fan and three are opened near the closed end (Case 5B).

When the layout of the exhaust vent is Case "3A", the smoke upstream can be controlled within 50 m of the last smoke vent; when the exhaust volume exceeds 220 m^3/s, the exhaust volume has little effect on smoke spread upstream. However, exhaust volume has a great influence on smoke spread downstream, and the length of the smoke spread decreases with the increase in exhaust rate, as shown in Figure 5a. When the layout of the smoke vent is Case "3B", the effect of exhaust volume on smoke spread is opposite to that of Case "3A"; the exhaust volume mainly affects the smoke spread upstream, as shown in Figure 5b. Moreover, the total length of the smoke spread (the sum of the spread lengths of both sides of the fire) in the tunnel for Case "3A" is significantly longer than that of Case "3B" at the same exhaust rate, as shown in Figure 6. Therefore, if only three smoke vents near the fire source are opened, the smoke control effect of Case "3B" is better than that of Case "3A".

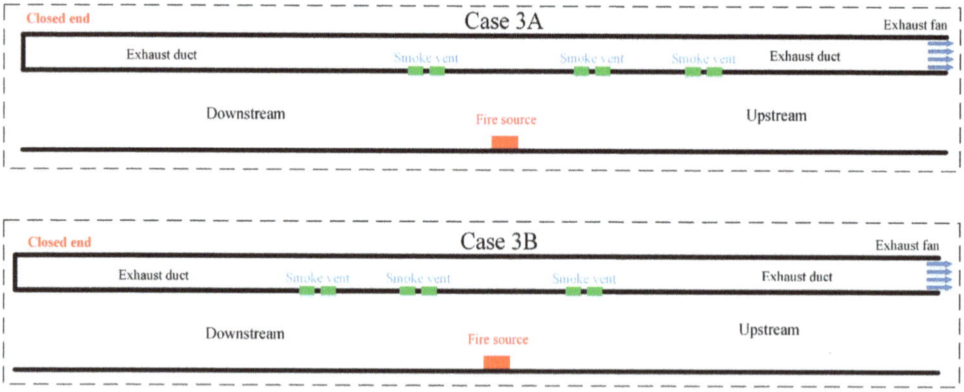

Figure 4. Layout of different smoke vent positions when 3 smoke vents are opened.

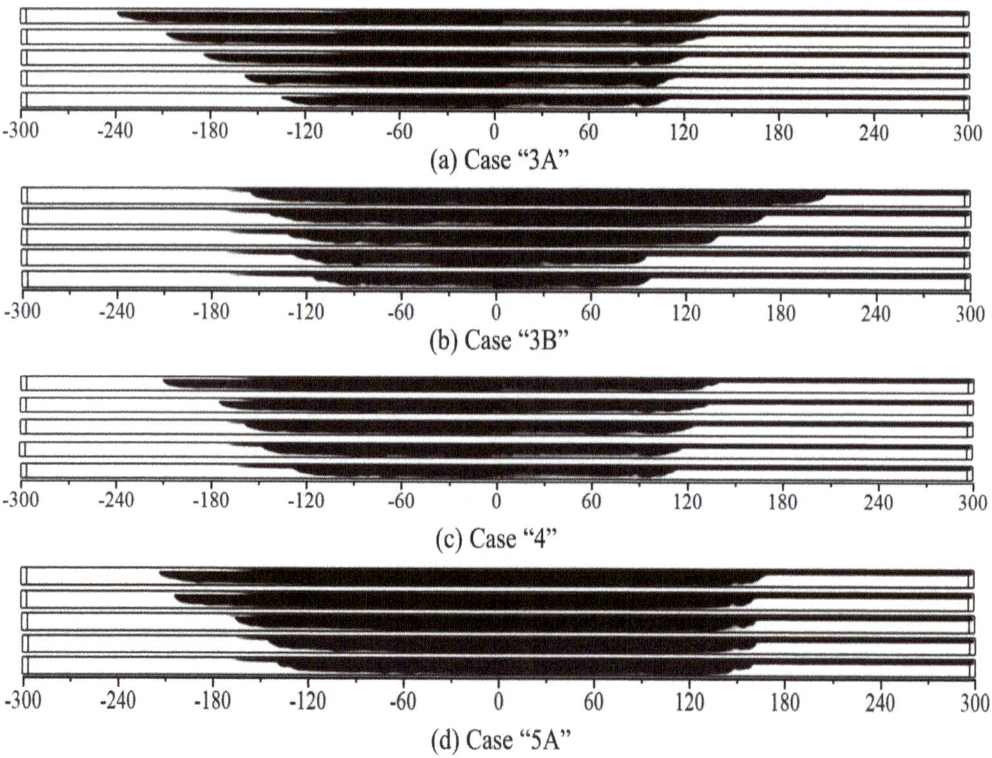

Figure 5. *Cont.*

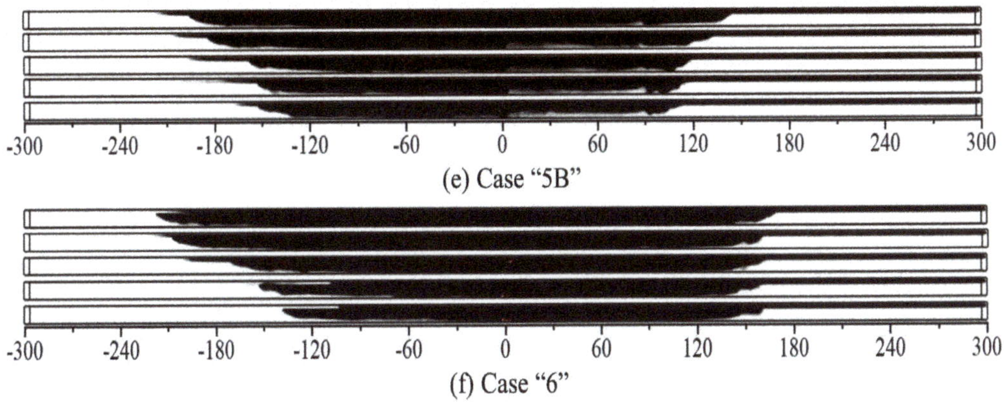

Figure 5. Smoke distribution with different number of smoke vents and exhaust rate (under each smoke vent scheme, the smoke exhaust rate is increased from 200 to 280 m^3/s).

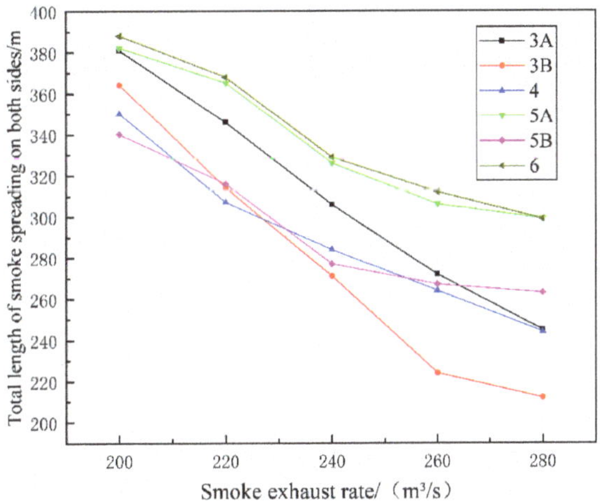

Figure 6. Smoke spreading length on both sides under different exhaust volume.

When four smoke vents are opened, the smoke distribution on both sides of fire is basically symmetrical. When the exhaust volume is more than 220 m^3/s, increasing the exhaust volume has no obvious effect on restraining the smoke spread, especially upstream. The smoke can be controlled within 150 m on both sides of the fire. When there are three smoke vents upstream and two downstream (Case 5A), the smoke spread length upstream increases obviously compared with Case "4". When the layout of the exhaust vent is Case "5B", the smoke spread length upstream decreases compared with Case "5A", as shown in Figure 5d,e. The influence of setting an exhaust fan on one side of the tunnel is highlighted. Due to the single-side point exhaust, the velocity of the third exhaust vent upstream is very high, which will help the smoke to spread to the third smoke vent. The smoke control effect of Case "5B" upstream is better than that of Case "5A".

When six smoke vents are opened, the smoke spreading length is the longest among all the smoke exhaust opening schemes. Most importantly, when the smoke exhaust rate is greater than 240 m^3/s, the smoke downstream cannot spread to the last smoke vent, which will cause serious waste. When three smoke vents are opened upstream, the smoke spread length is much longer than when opening one or two smoke vents upstream, and the total smoke spread length in the tunnel exceeds 300 m (the longest is 388 m) under different exhaust rates. The total length of the smoke spread is more than 200 m in all calculation conditions. From the perspective of smoke diffusion, for a single-side point exhaust tunnel, the best smoke exhaust effect can be obtained when four exhaust vents are opened, especially for a two-way tunnel.

3.2. *Temperature Distribution*

3.2.1. Maximum Temperature

When the layout of the smoke vents is Case "5A" and Case "6", the maximum temperature decreases with the increase in exhaust volume; the maximum temperature under the tunnel ceiling is 719–918 °C, as shown in Figure 7. When the layout of the exhaust vent is Case "2", Case "3A", Case "3B", Case "4", and Case "5B", the maximum temperature first increases and then decreases with the increase in exhaust volume. This is because when the smoke is confined near the fire source and sinks to the bottom of tunnel, the fire is wrapped by the smoke and does not burn sufficiently. Increasing the exhaust volume strengthens the air convection in the tunnel and makes the maximum temperature rise. If the exhaust volume continues to increase, the velocity at the vent nearest to the fire will make the high-temperature smoke unable to gather in the vault, and the smoke vent far away from the fire will restrict air convection, which will reduce the maximum temperature. When the layout of the exhaust vent is Case "5A" and Case "6", the smoke within the range of the smoke vents will sink to the bottom of the tunnel, as shown in Figure 4. It is difficult for fresh air to reach the fire source with the increase in exhaust volume, so the maximum temperature will not increase suddenly.

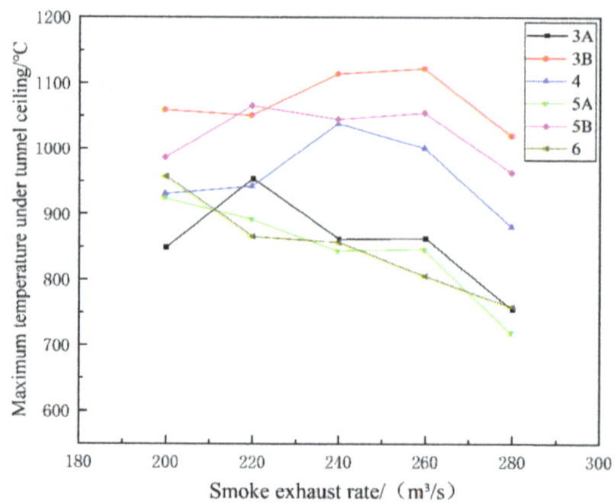

Figure 7. Maximum temperature with different exhaust volume.

When the layout of the exhaust vent is Case "4", the exhaust volume has a great influence on the maximum temperature. When the exhaust volume increases from 220 m^3/s to 240 m^3/s, the maximum temperature increases by 97 °C. Although the maximum temperature decreases rapidly with the increase in the exhaust volume, it is still much higher than that of the other schemes. When the exhaust volume is less than 220 m^3/s, the maximum temperature is almost the same as that of the other schemes. If only four smoke vents are opened, the exhaust rate should not be more than 220 m^3/s.

3.2.2. Temperature Decay Model

For the downstream smoke under ceilings far away from fire source:

Mass equation:

$$\frac{\mathrm{d}}{\mathrm{d}x}(\rho u A) = \rho_a W u_e \tag{2}$$

Energy equation:

$$\frac{\mathrm{d}}{\mathrm{d}x}(\rho A u c_p T) = \rho_a W u_e c_p T_a - h_t w_p (T - T_a) \tag{3}$$

The entrainment velocity u_e can be expressed as:

$$u_e = \beta(u - u_o) \tag{4}$$

where ρ and u are the density (kg/m^3) and velocity (m/s) of smoke; A is the smoke flow section area (m^2); W is the width of the tunnel (m); T is the temperature of the smoke (K); u_e is the entrainment velocity of the smoke (m/s); w_p is the wet perimeter of the smoke flow (m); h_t is the total net heat transfer coefficient on the tunnel walls (kW/m^2·K); β is a coefficient measured in the test; and u_o is the velocity of the air (m/s).

In tunnel fires, smoke entrainment velocity is very small [41]. In order to simplify the energy equation, we usually ignore the influence of smoke entrainment on the energy equation. The tunnel wet perimeter can be calculated according to the tunnel design parameters, so the tunnel wet perimeter can be considered a constant. It is assumed that h_t is also a constant. Based on Equations (2)–(4), the temperature attenuation downstream is obtained as:

$$\frac{\Delta T(x)}{\Delta T_{\max}} = \exp\left(-\frac{h_t w_p + \rho_a u_e W c_p}{\rho u A c_p} x\right) \tag{5}$$

Because the coefficient β (=0.00015) is very small [42], the horizontal entrainment of smoke in the tunnel can be ignored:

$$\frac{\Delta T(x)}{\Delta T_{\max}} \approx \exp\left(-\frac{h_t w_p}{\rho u A c_p} x\right) \tag{6}$$

For a tunnel with a rectangular section, the temperature decay downstream can be expressed as:

$$\frac{\Delta T(x)}{\Delta T_{\max}} \approx \exp\left(-(\frac{2h}{W}+1)\cdot\frac{h_t}{\rho u c_p}\cdot\frac{x}{h}\right) = \exp\left(-\xi\cdot\frac{x}{h}\right) \tag{7}$$

$$\xi = (\frac{2h}{W}+1)\cdot\frac{h_t}{\rho u c_p} \propto \frac{h_t}{\rho u c_p} \tag{8}$$

Although the coefficient h_t is different in different locations and different fire development stages, when the fire tends to be stable and far enough from fire, the difference of h_t in different locations is very small. In addition, the height of the smoke layer can also be replaced by tunnel height. In view of this, the temperature decay can be approximated using an exponential function [20,40]. Li et al. [43] found that the form of the sum of two exponential functions, that is $\Delta T(x)/\Delta T_{\max} = a \times \exp\left(b\times\frac{x}{H}\right) + c \times \exp\left(d\times\frac{x}{H}\right)$, can well describe the decay of temperature under the tunnel ceiling.

Taking the location of the maximum temperature as a reference point, this paper studies the temperature decay from the reference point to the first vent downstream. The smoke control effect is the best in Case "4", and the decay rate of the temperature is like that of other smoke vent opening schemes. Therefore, this paper only presents the fitting results for the temperature attenuation of Case "4".

When four smoke vents are opened, the temperature decay rate is the same under different smoke exhaust rates (H is the height of the tunnel, m), as shown in Figure 8. The results of numerical calculation are in good agreement with the fitting curve (R^2 = 0.992). The constants a, b, c, and d are 0.444, −0.156, 0.598, and −1.913, respectively. The temperature decays the slowest when the layout of the exhaust vent is Case "5B" and decays the fastest when the layout of exhaust vent is Case "3B". The correlation coefficient (R^2) of the fitting curve of the temperature attenuation under different smoke vent opening schemes was greater than 0.976. Except for the smoke vent opening schemes of Case "5B" and Case

"3A", the temperature decays rapidly and the decay rates are similar. Furthermore, the fitting curves of the scheme with fast decay rate are close to the results of temperature for Case "4". Therefore, we only need to focus on the temperature attenuation of Case "4".

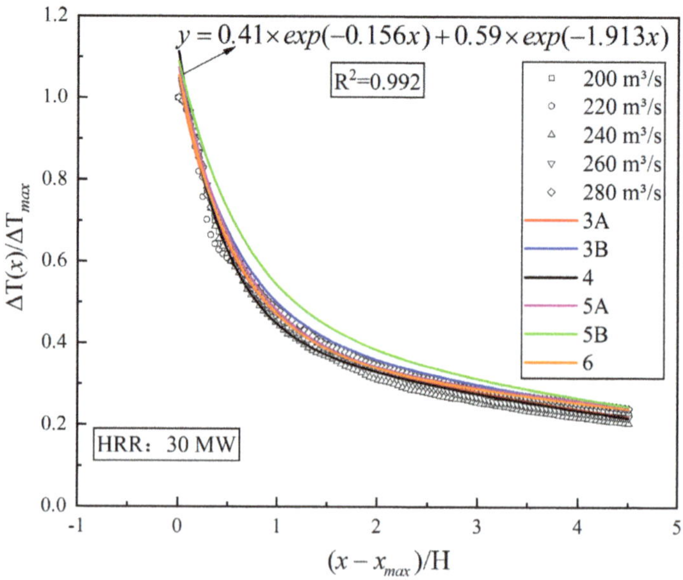

Figure 8. Temperature attenuation beneath tunnel ceiling of Case "4".

In order to make the empirical formula for the temperature attenuation more representative, the temperature distributions under different HRRs and exhaust volumes are calculated for Case "4". The detailed calculation conditions are shown in Table 4. Under different HRRs, the exhaust volume has little effect on the temperature attenuation, as shown in Figure 9. The influence of the HRR on the attenuation rate is very small. When the HRR of the fire source is 10–30 MW, the temperature decay rate is similar. Taking the average value of each coefficient, the empirical formula of temperature attenuation under the tunnel ceiling with four smoke vents open can be obtained as follows:

$$\Delta T(x)/\Delta T_{max} = 0.40e^{-0.147(\frac{x-x_{max}}{H})} + 0.60e^{-2.17(\frac{x-x_{max}}{H})} \quad (9)$$

Table 4. Calculation condition of 4 smoke vents.

Test No.	Smoke Vent Number	HRR (MW)	Smoke Exhaust Rate (m^3/s)
31–35	4	10	100, 120, 140, 160, 180
36–40		20	160, 180, 200, 220, 240

To highlight the difference between the single-side multi-point exhaust tunnel and the previous research results, we compared the temperature attenuation model in this paper with some existing temperature attenuation models. Ji et al. [37] studied the effect of pressure and HRR on the temperature decay beneath the ceiling and found that the impact of pressure on temperature decay is very small, and the temperature attenuation conforms to the sum of two exponential attenuations:

$$\Delta T_x/\Delta T_r = 0.33e^{-0.59\frac{(x-x_r)}{H}} + 0.67e^{-0.048\frac{(x-x_r)}{H}} \quad (10)$$

Ingason and Li [43] concluded that the temperature attenuation beneath the ceiling conforms to the sum of two exponential attenuations:

$$\Delta T_x / \Delta T_r = 0.57e^{-0.13\frac{(x-x_r)}{H}} + 0.43e^{-0.021\frac{(x-x_r)}{H}} \quad (11)$$

The numerical results are in good agreement with the prediction model proposed in this paper, and the error between the numerical results and the prediction model is basically within 15%, as shown in Figure 10. The closer to the smoke vent, the greater the error between the numerical results and the prediction model, with the numerical calculation results slightly higher than the prediction model. This is because the smoke vent will inhibit the movement of high-temperature smoke and reduce the temperature attenuation rate. Because there are four smoke vents in the tunnel that are continuously discharging high-temperature smoke, the rate of the temperature attenuation model in this paper is faster than that of natural ventilation or longitudinal ventilation tunnel, and the choice of reference points may also be one of the reasons for the large differences.

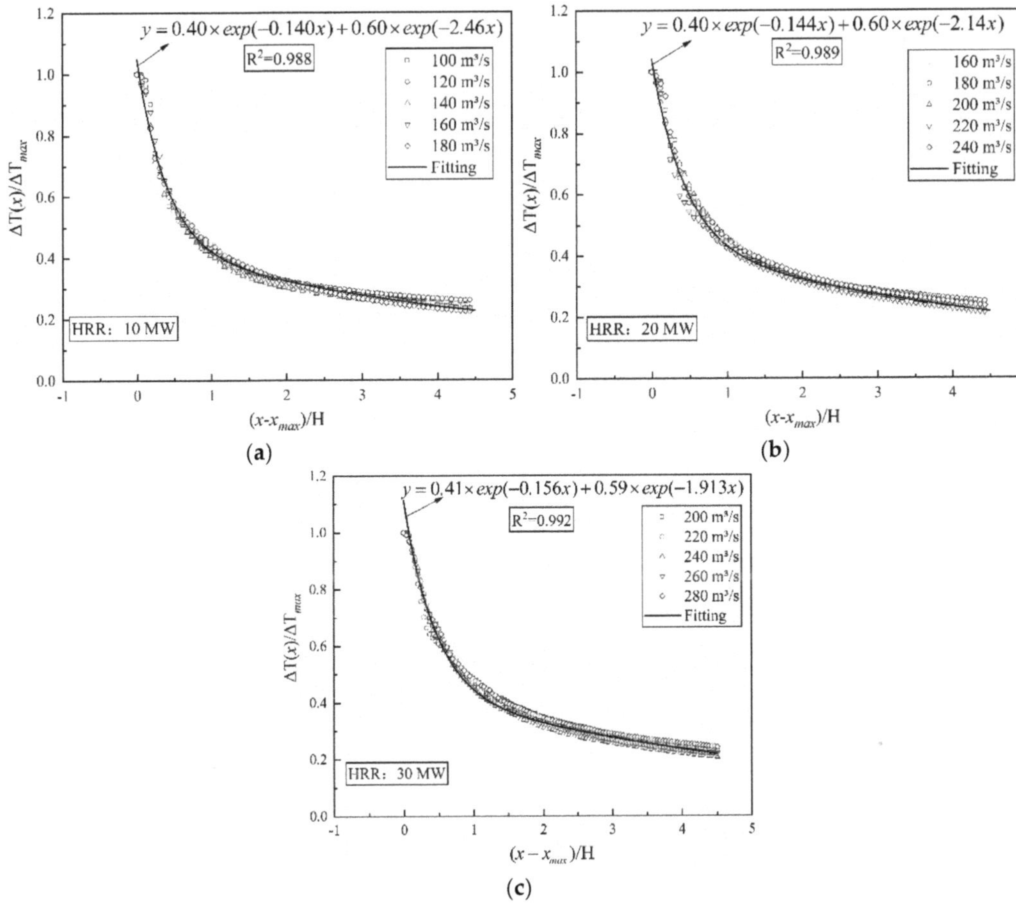

Figure 9. Temperature attenuation with different HRRs of Case "4". (**a**) 10 MW, (**b**) 20 MW, (**c**) 30 MW.

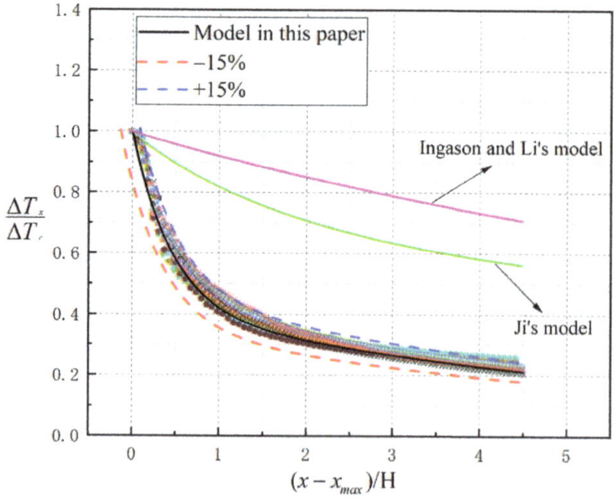

Figure 10. Comparison between existing temperature decay models and the model proposed in this paper.

3.3. Exhaust Efficiency

By comparing the mass flow rate of CO in the exhaust duct under different calculation conditions, the efficiency of smoke extraction with different calculation conditions can be judged [30]. When the layout of the smoke vent is Case "4", the increase in total smoke exhaust efficiency is very small with increasing exhaust volume, and the effect of the exhaust rate on total smoke exhaust efficiency is very small, as shown in Figure 11. Under different exhaust volumes, the total exhaust efficiency is more than 93.7% when the layout of the exhaust vent is Case "4". The maximum difference of total smoke exhaust efficiency is less than 1.5% under different exhaust volumes in Case "4". Moreover, the total exhaust efficiency with the same exhaust rate is significantly higher than that of other smoke vent opening schemes. When the layout of the exhaust vent is Case "3A", the total smoke exhaust efficiency is the lowest compared with other schemes, and the smoke exhaust efficiency is greatly affected by the exhaust rate. When the exhaust volume increases by 20 m^3/s, the total smoke exhaust efficiency increases by more than 2.5%.

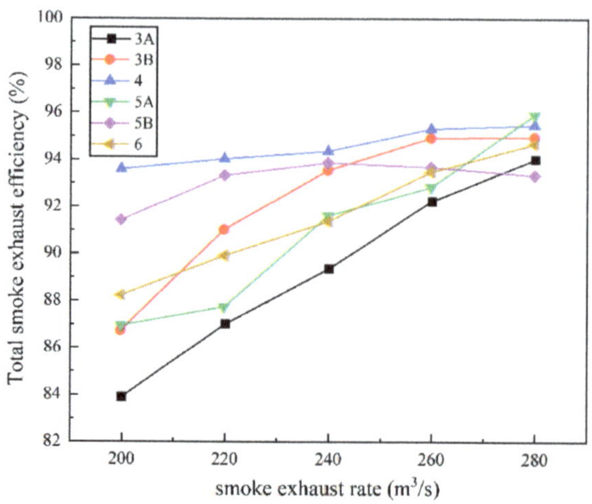

Figure 11. Total smoke exhaust efficiency under different exhaust rates.

The trend of total exhaust efficiency under different smoke vent opening states is that the higher the exhaust rate, the greater the total exhaust efficiency. However, if the exhaust rate exceeds 240 m^3/s, the total smoke exhaust efficiency decreases with the increase in exhaust volume in Case "5B". This is because the velocity at the smoke vent downstream is far less than that upstream, and the farther away from the fire, the lower the velocity at the smoke vent. As the exhaust volume increases, the smoke may not spread to the smoke vent, which will result in the intake of fresh air instead of smoke. It can be noticed from Figure 12c,d that when the exhaust volume exceeds 260 m^3/s, the exhaust efficiency of the smoke vent farthest downstream from the fire is almost 0%.

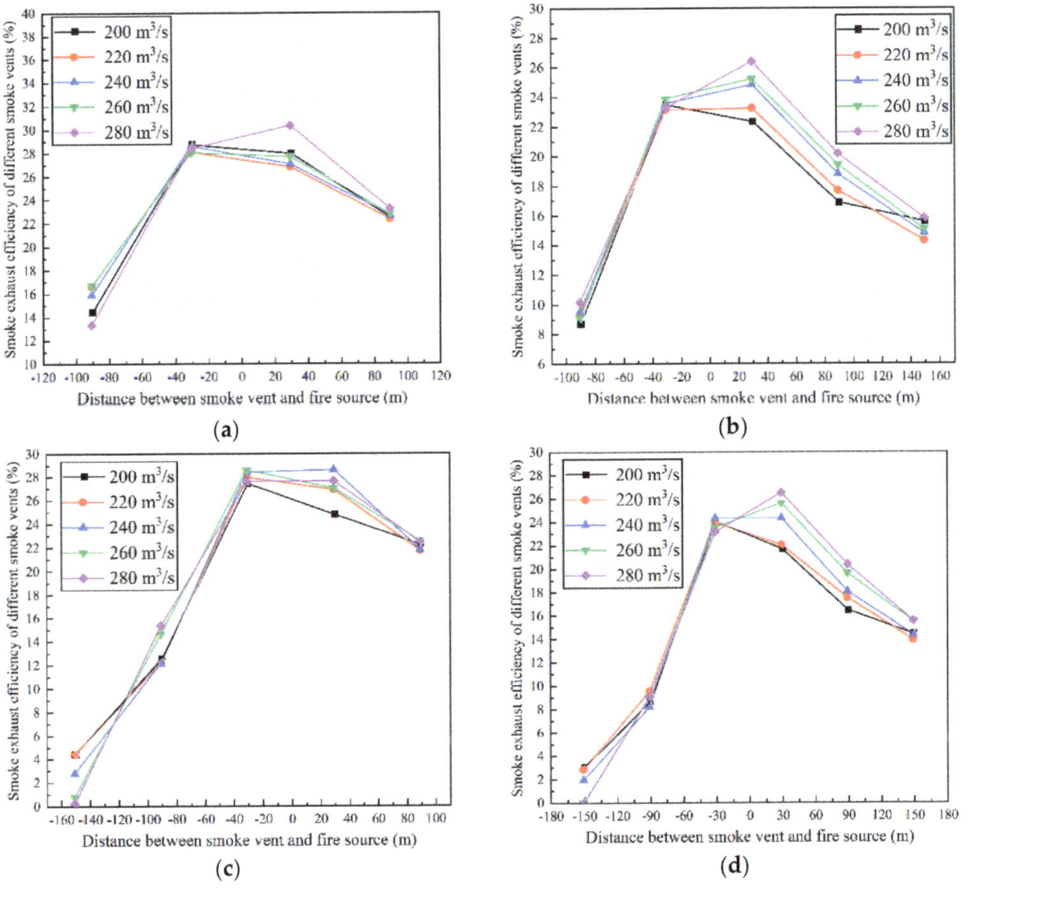

Figure 12. Efficiency of each exhaust vent under different exhaust rates. (**a**) Case "4", (**b**) Case "5A", (**c**) Case "5B", (**d**) Case "6".

With the increase in distance from the fire, the smoke exhaust efficiency of the smoke vent downstream decreases rapidly, as shown in Figure 12. The smoke vent with the highest smoke exhaust efficiency is the nearest one on both sides of the fire. In the process of smoke spreading, the smoke will continuously entrain the fresh air in the tunnel, resulting in the decrease in CO content. Therefore, the CO content near the fire source is the highest. For the smoke vent upstream, the farther away from fire, the higher the exhaust velocity, the lower the exhaust efficiency. Moreover, the exhaust volume has little impact on the exhaust efficiency of the vent closest to the exhaust fan. The smoke exhaust efficiency of the other smoke vents upstream increases with the exhaust rate. The smoke is mainly discharged from the two smoke vents nearest to the fire source, and the exhaust volume has little impact on the exhaust efficiency (the maximum difference of the exhaust efficiency with different exhaust rates is less than 4%). When three smoke vents are opened downstream, the smoke vent 150 m away from the fire makes little contribution to controlling the smoke, and the smoke exhaust efficiency is less than 5% under different smoke exhaust rates.

4. Conclusions

A set of CFD simulations were performed in single-side point exhaust tunnels with different smoke exhaust rates to explore the impact of the number of smoke vents on smoke control, temperature profile, and smoke exhaust efficiency. The main conclusions are as follows:

(1) When there are more than two smoke vents on one side of the fire source far away from the exhaust fan, some smoke exhaust vents will inhale fresh air rather than toxic smoke. When the layout of the exhaust vent is Case "4", the total length of the smoke spread in single-side point exhaust tunnels is shortest;

(2) The maximum temperature decreases with the increase in the smoke exhaust rate. The temperature is much higher than in other smoke exhaust vent opening schemes at the same smoke exhaust volume as Case "6";

(3) By analyzing the simulation results of the vault temperature under different HRRs and exhaust rates, an empirical formula of temperature attenuation for Case "4" was proposed: $\Delta T(x)/\Delta T_{max} = 0.40e^{-0.147(\frac{x-x_{max}}{H})} + 0.60e^{-2.17(\frac{x-x_{max}}{H})}$. The error of the temperature attenuation model is less than 15%;

(4) Under the same exhaust volume, the exhaust efficiency is the highest when the layout of the exhaust vent is Case "4". The total smoke exhaust efficiency of the tunnel is more than 93.7% and the maximum difference of the total smoke exhaust efficiency is less than 1.5% under different smoke exhaust rates;

(5) For a single-side point exhaust tunnel, the number of smoke vents near the smoke exhaust fan side shall not be more than that on the other side. The proposed smoke control scheme for the Mawan tunnel with a designed HRR of 30 MW is as follows: the layout of the exhaust vent is as shown in Case "4", and the smoke exhaust rate is 220 m^3/s.

This paper mainly studied the smoke and temperature control of a centralized smoke exhaust tunnel, discussed the smoke exhaust effect when there are different numbers of smoke outlets, solved the fire ventilation design of Mawan tunnel, and provided a reference for a similar tunnel fire ventilation design. A study on the temperature field of a single-side point exhaust tunnel fire, especially the maximum temperature, will be carried out in further model tests.

Author Contributions: Conceptualization, L.T. and Y.Z.; methodology, L.T. and Y.Z.; software, L.T.; validation, L.T.; formal analysis, L.T. and Y.Z.; data curation, L.T.; writing—original draft preparation, L.T. and Y.Z.; writing—original draft preparation, L.T. and Y.Z.; writing—review and editing, L.T. and Y.Z.; project administration, Y.Z.; funding acquisition, Y.Z.; All authors have read and agreed to the published version of the manuscript.

Funding: This work was supported by the National Natural Foundation of China (No. 52178394).

Institutional Review Board Statement: Not applicable.

Informed Consent Statement: Not applicable.

Data Availability Statement: Not applicable.

Acknowledgments: The authors greatly appreciate the financial support of the Shenzhen Transportation Public Facilities Construction Center.

Conflicts of Interest: The authors declare no potential conflict of interest with respect to the research, authorship, and/or publication of this article.

References

1. Tomar, M.S.; Khurana, S. Impact of passive fire protection on heat release rates in road tunnel fire: A review. *Tunn. Undergr. Space Technol.* **2019**, *85*, 149–159. [CrossRef]
2. Hua, N.; Elhami Khorasani, N.; Tessari, A.; Ranade, R. Experimental study of fire damage to reinforced concrete tunnel slabs. *Fire Saf. J.* **2022**, *127*, 103504. [CrossRef]
3. Felis, F.; Pavageau, M.; Elicer-Cortés, J.C.; Dassonville, T. Simultaneous measurements of temperature and velocity fluctuations in a double stream-twin jet air curtain for heat confinement in case of tunnel fire. *Int. Commun. Heat Mass Transf.* **2010**, *37*, 1191–1196. [CrossRef]
4. Alpert, R.L. Calculation of response time of ceiling-mounted fire detectors. *Fire Technol.* **1972**, *8*, 181–195. [CrossRef]
5. Alarie, Y. Toxicity of fire smoke. *Crit. Rev. Toxicol.* **2002**, *32*, 259–289. [CrossRef]
6. Tao, L.; Zhang, Y.; Hou, K.; Bai, Y.; Zeng, Y.; Fang, Y. Experimental study on temperature distribution and smoke control in emergency rescue stations of a slope railway tunnel with semi-transverse ventilation. *Tunn. Undergr. Space Technol.* **2020**, *106*, 103616. [CrossRef]
7. Vauquelin, O. Experimental simulations of fire-induced smoke control in tunnels using an "air–helium reduced scale model": Principle, limitations, results and future. *Tunn. Undergr. Space Technol.* **2008**, *23*, 171–178. [CrossRef]
8. Lönnermark, A.; Ingason, H. Gas temperatures in heavy goods vehicle fires in tunnels. *Fire Saf. J.* **2005**, *40*, 506–527. [CrossRef]
9. Lönnermark, A.; Ingason, H. Fire Spread and Flame Length in Large-Scale Tunnel Fires. *Fire Technol.* **2006**, *42*, 283–302. [CrossRef]
10. Li, M.; Qiang, Y.; Wang, X.; Shi, W.; Zhou, Y.; Yi, L. Effect of Wind Speed on the Natural Ventilation and Smoke Exhaust Performance of an Optimized Unpowered Ventilator. *Fire* **2022**, *5*, 18. [CrossRef]
11. Zhao, P.; Yuan, Z.; Yuan, Y.; Yu, N.; Yu, T. A Study on Ceiling Temperature Distribution and Critical Exhaust Volumetric Flow Rate in a Long-Distance Subway Tunnel Fire with a Two-Point Extraction Ventilation System. *Energies* **2019**, *12*, 1411. [CrossRef]
12. Tang, F.; Mei, F.Z.; Wang, Q.; He, Z.; Fan, C.G.; Tao, C.F. Maximum temperature beneath the ceiling in tunnel fires with combination of ceiling mechanical smoke extraction and longitudinal ventilation. *Tunn. Undergr. Space Technol.* **2017**, *68*, 231–237. [CrossRef]
13. Tang, F.; Chen, L.; Chen, Y.; Pang, H. Experimental study on the effect of ceiling mechanical smoke extraction system on transverse temperature decay induced by ceiling jet in the tunnel. *Int. J. Therm. Sci.* **2020**, *152*, 106294. [CrossRef]
14. Tang, F.; Li, L.J.; Mei, F.Z.; Dong, M.S. Thermal smoke back-layering flow length with ceiling extraction at upstream side of fire source in a longitudinal ventilated tunnel. *Appl. Therm. Eng.* **2016**, *106*, 125–130. [CrossRef]
15. Zhu, Y.; Tang, F.; Chen, L.; Wang, Q.; Xu, X. Effect of lateral concentrated smoke extraction on the smoke back-layering length and critical velocity in a longitudinal ventilation tunnel. *J. Wind Eng. Ind. Aerodyn.* **2020**, *207*, 104403. [CrossRef]
16. Wang, J.; Yuan, J.; Fang, Z.; Tang, Z.; Qian, P.; Ye, J. A model for predicting smoke back-layering length in tunnel fires with the combination of longitudinal ventilation and point extraction ventilation in the roof. *Tunn. Undergr. Space Technol.* **2018**, *80*, 16–25. [CrossRef]
17. He, L.; Xu, Z.; Chen, H.; Liu, Q.; Wang, Y.; Zhou, Y. Analysis of entrainment phenomenon near mechanical exhaust vent and a prediction model for smoke temperature in tunnel fire. *Tunn. Undergr. Space Technol.* **2018**, *80*, 143–150. [CrossRef]
18. Jiang, X.; Liu, M.; Wang, J.; Li, K. Study on air entrainment coefficient of one-dimensional horizontal movement stage of tunnel fire smoke in top central exhaust. *Tunn. Undergr. Space Technol.* **2016**, *60*, 1–9. [CrossRef]
19. Tao, L.; Zeng, Y. Effect of single-side centralized exhaust on smoke control and temperature distribution in longitudinal ventilation tunnel fires. *Tunn. Undergr. Space Technol.* **2022**, *119*, 104241. [CrossRef]
20. Tao, L.; Zeng, Y.; Li, J.; Yang, G.; Fang, Y.; Li, B. Study on the maximum temperature and temperature decay in single-side centralized smoke exhaust tunnel fires. *Int. J. Therm. Sci.* **2022**, *172*, 107277. [CrossRef]
21. Yan, Z.; Zhang, Y.; Guo, Q.; Zhu, H.; Shen, Y.; Guo, Q. Numerical study on the smoke control using point extraction strategy in a large cross-section tunnel in fire. *Tunn. Undergr. Space Technol.* **2018**, *82*, 455–467. [CrossRef]
22. Xu, Z.; Yangyang, L.; Rui, K.; Junpeng, G.; Long, Y. Study on the Reasonable Smoke Exhaust Rate of the Crossrange Exhaust Duct in Double-layer Shield Tunnel. *Procedia Eng.* **2014**, *84*, 506–513. [CrossRef]
23. Vauquelin, O.; Mégret, O. Smoke extraction experiments in case of fire in a tunnel. *Fire Saf. J.* **2002**, *37*, 525–533. [CrossRef]
24. Blanchard, E.; Boulet, P.; Desanghere, S.; Cesmat, E.; Meyrand, R.; Garo, J.P.; Vantelon, J.P. Experimental and numerical study of fire in a midscale test tunnel. *Fire Saf. J.* **2012**, *47*, 18–31. [CrossRef]
25. Lin, C.J.; Chuah, Y.K. A study on long tunnel smoke extraction strategies by numerical simulation. *Tunn. Undergr. Space Technol.* **2008**, *23*, 522–530. [CrossRef]
26. Gannouni, S.; Maad, R.B. Numerical study of the effect of blockage on critical velocity and backlayering length in longitudinally ventilated tunnel fires. *Tunn. Undergr. Space Technol.* **2015**, *48*, 147–155. [CrossRef]
27. Tanaka, F.; Kawabata, N.; Ura, F. Smoke spreading characteristics during a fire in a shallow urban road tunnel with roof openings under a longitudinal external wind blowing. *Fire Saf. J.* **2017**, *90*, 156–168. [CrossRef]
28. Halawa, T.; Safwat, H. Fire-smoke control strategies in road tunnels: The effectiveness of solid barriers. *Case Stud. Therm. Eng.* **2021**, *27*, 101260. [CrossRef]
29. Babrauskas, V. Estimating large pool fire burning rates. *Fire Technol.* **1983**, *19*, 251–261. [CrossRef]
30. Yao, Y.; Li, Y.Z.; Ingason, H.; Cheng, X. Numerical study on overall smoke control using naturally ventilated shafts during fires in a road tunnel. *Int. J. Therm. Sci.* **2019**, *140*, 491–504. [CrossRef]
31. Baum, H.R.; Mccaffrey, B.J. Fire Induced Flow Field—Theory and Experiment. *Fire Saf. Sci. Proc. Second Int. Symp.* **1989**, *2*, 129–148. [CrossRef]

32. Ji, J.; Guo, F.; Gao, Z.; Zhu, J. Effects of ambient pressure on transport characteristics of thermal-driven smoke flow in a tunnel. *Int. J. Therm. Sci.* **2018**, *125*, 210–217. [CrossRef]
33. Ji, J.; Tan, T.; Gao, Z.; Wan, H.; Zhu, J.; Ding, L. Numerical Investigation on the Influence of Length–Width Ratio of Fire Source on the Smoke Movement and Temperature Distribution in Tunnel Fires. *Fire Technol.* **2019**, *55*, 963–979. [CrossRef]
34. Ji, J.; Wang, Z.; Ding, L.; Yu, L.; Gao, Z.; Wan, H. Effects of ambient pressure on smoke movement and temperature distribution in inclined tunnel fires. *Int. J. Therm. Sci.* **2019**, *145*, 106006. [CrossRef]
35. Liu, C.; Zhong, M.; Song, S.; Xia, F.; Tian, X.; Yang, Y.; Long, Z. Experimental and numerical study on critical ventilation velocity for confining fire smoke in metro connected tunnel. *Tunn. Undergr. Space Technol.* **2020**, *97*, 103296. [CrossRef]
36. Liang, Q.; Li, Y.; Li, J.; Xu, H.; Li, K. Numerical studies on the smoke control by water mist screens with transverse ventilation in tunnel fires. *Tunn. Undergr. Space Technol.* **2017**, *64*, 177–183. [CrossRef]
37. Chen, L.; Du, S.; Zhang, Y.; Xie, W.; Zhang, K. Experimental study on the maximum temperature and flame extension length driven by strong plume in a longitudinal ventilated tunnel. *Exp. Therm. Fluid Sci.* **2019**, *101*, 296–303. [CrossRef]
38. Tao, L.; Yan, X.; Zhang, Y.; Zeng, Y.; Fang, Y.; Bai, Y. Experimental and numerical study on the smoke and velocity distribution in an extra-long railway tunnel fire. *Tunn. Undergr. Space Technol.* **2021**, *117*, 104134. [CrossRef]
39. Hu, L.H.; Huo, R.; Chow, W.K. Studies on buoyancy-driven back-layering flow in tunnel fires. *Exp. Therm. Fluid Sci.* **2008**, *32*, 1468–1483. [CrossRef]
40. Kunsch, J.P. Critical velocity and range of a fire-gas plume in a ventilated tunnel. *Atmos. Environ.* **1999**, *33*, 13–24. [CrossRef]
41. Ingason, H.; Li, Y.Z. Model scale tunnel fire tests with longitudinal ventilation. *Fire Saf. J.* **2010**, *45*, 371–384. [CrossRef]
42. Ji, J.; Guo, F.Y.; Gao, Z.H.; Zhu, J.P.; Sun, J.H. Numerical investigation on the effect of ambient pressure on smoke movement and temperature distribution in tunnel fires. *Appl. Therm. Eng.* **2017**, *118*, 663–669. [CrossRef]
43. Guo, F.; Gao, Z.; Wan, H.; Ji, J.; Yu, L.; Ding, L. Influence of ambient pressure on critical ventilation velocity and backlayering distance of thermal driven smoke in tunnels with longitudinal ventilation. *Int. J. Therm. Sci.* **2019**, *145*, 105989. [CrossRef]

fire

Article

Experimental and Theoretical Analysis of the Smoke Layer Height in the Engine Room under the Forced Air Condition

Xiaowei Wu [1,2], Yi Zhang [1], Jia Jia [2], Xiao Chen [1,*], Wenbing Yao [1] and Shouxiang Lu [1,*]

1 State Key Laboratory of Fire Science, University of Science and Technology of China, Hefei 230026, China
2 Naval Research Institute, Beijing 100161, China
* Correspondence: summercx@ustc.edu.cn (X.C.); sxlu@ustc.edu.cn (S.L.); Tel.: +86-187-2551-5302 (X.C.); +86-133-6551-9099 (S.L.)

Abstract: The smoke layer height in the ship engine room under forced ventilation has been experimental and theoretical investigated in this work. A series of test were carried out in a scaled engine cabin experimental platform to obtain the influence of air supply volume and air inlet height on the burning parameters, including the mass loss rate, smoke temperature, etc. The research results show that under the experimental conditions, the fire source mass loss rate increases exponentially, and smoke layer height also increases gradually with the increase in the air supply volume. The empirical formula of smoke layer height under different air supply conditions was given. Then, a prediction model of smoke layer height under different forced ventilation conditions was constructed through theoretical analysis based on conservation equations. Within the range of experimental air volume and air inlet height, the relative error between theoretical prediction results and experimental results was less than 11%, which could effectively predict the smoke layer height in the ship cabin fire.

Keywords: engine room; fire; forced air supply; smoke layer height

Citation: Wu, X.; Zhang, Y.; Jia, J.; Chen, X.; Yao, W.; Lu, S. Experimental and Theoretical Analysis of the Smoke Layer Height in the Engine Room under the Forced Air Condition. *Fire* **2023**, *6*, 16. https://doi.org/10.3390/fire6010016

Academic Editors: Chuangang Fan and Dahai Qi

Received: 4 November 2022
Revised: 27 December 2022
Accepted: 30 December 2022
Published: 4 January 2023

1. Introduction

Ship fire accidents often cause serious damage to personnel and other important equipment on board, making ship fires a major threat to ship safety. Among all kinds of ship fire accidents, engine room fire is the main research object at present because of the characteristics of high fire frequency, large fire load, and high difficulty in fire detection and rescue.

In recent years, the problem of smoke control in ship fires has attracted much attention [1–4]. The smoke control of the engine room mainly relies on the mechanical smoke exhaust system and the mechanical air supply system. When fire occurs, the smoke control system is activated, and the original smoke flow of the engine room will change. Therefore, the law of cabin smoke filling under forced ventilation conditions is of great significance for the study of the fire hazard of the engine room.

Previous studies have attracted more attention to the effects of forced ventilation on the development of cabin fires and the law of smoke filling in closed cabins. Many scholars have carried out closed space fires under different forced ventilation conditions, and have studied the effect of forced ventilation on fire parameters such as fire source burning rate [5–8], temperature distribution [9–12], and cabin pressure [13–15]. Alvares [5] found similar fire parameters for two different tested liquid fuels under the same fire intensity and ventilation. Chow and Chan [6] provided more information on the effect of fuel type and found that fire parameters for wood were sensitive to changes in ventilation volume, whereas fire parameters for liquid fuels and polymer materials were less sensitive to changes in ventilation volume. Peatross and Beyler [7] believed that the burning rate was determined by the fuel type and ventilation rate, and the forced ventilation test results showed that the burning rate increased with the increase in ventilation rate. Le [8] analyzed that, in a mechanically ventilated compartment, the mass loss rate of ignition source depends on two factors: the oxygen concentration and the blowing effect on the oil pool. In addition, different ventilation configurations, such as ventilation volume, and vent height, can have an impact on combustion characteristics in the cabin. Backovsky [16] studied the effect of ventilation volume and ventilation configuration on the chamber temperature, and the results showed that under the condition of low air inlet position, the ventilation rate was 2–3 times greater than the stoichiometric air volume required for fires (here called well-ventilated fires) produce a two-layer temperature distribution, whereas under-ventilated fires produce a single-layer distribution with a temperature gradient, with poorly ventilated fires having higher temperatures than well-ventilated fires. The high-inlet location perturbs the two-layer temperature distribution of a well-ventilated fire. Zhang [4] found that increasing the height of the air inlet or reducing the height of the air inlet can destroy the formation of the smoke layer. Based on the conservation equation of oxygen concentration, Hayashi [17] considered the oxygen volume exchange between the upper and lower layers due to convection and turbulent diffusion, revealed the conditions for the formation of the smoke layer under forced ventilation, and studied the formation and destruction of smoke layer by CFD. In previous studies on the effect of forced ventilation on cabin fires, there are relatively few studies on the height of the smoke layer under forced ventilation conditions.

Smoke layer height is one of the important parameters in compartment fire research. Existing calculation methods for fire smoke layer height include the classical N-percentage, integral ratio, intra-variance methods and so on. These methods are mostly suitable for the processing of temperature data. Predecessors have carried out a lot of research on the change in smoke layer height in compartment fire [18–20]. Most of these studies are experimental and simulation studies based on the corresponding structures of buildings. For the study of the height of the smoke layer, some common models have been formed, such as the NFPA 92B model [21], the Milke Mowrer model and the Tanaka [22] model. It has been more than 20 years since Zukoski [23] first proposed a simple analytical model of enclosed smoke filling based on energy and mass balance. The Zukoski formula remains the theoretical basis for other enclosed space smoke filling models, including the ASET model developed by Cooper. Li [24] obtained a formula for predicting the smoke layer based on the regional model (there is an obvious layering between the upper hot smoke and the lower cold air) and according to the conservation of energy. Li [25] carried out the natural gas filling experiment and the smoke control experiment in the experimental cabin, and based on the plume theory and the conservation equation, he developed the smoke settlement formula for the filled cabin, and considered the plume rising time. Wang [26] established an explicit smoke-filling model of an enclosed ship engine room through the back analysis of the experimental data in the literature.

It can be seen that there is a lack of theoretical and experimental research on the height of the smoke layer of closed cabin fires under forced ventilation in the current research on engine room fires. Therefore, this paper studies the development of the cabin room fire and the process of smoke settlement under the conditions of different air supply volume and air supply inlet heights through the small-scale engine room fire experiment. the theoretical analysis was carried out to predict the height of the cabin fire smoke layer un-der different air supply conditions based on the zone model, and the prediction results of the model are compared with the experimental results to verify the validity of the theoretical model. The model proposed in this paper can provide a theoretical analysis and prediction method for ship fire risk research.

2. Experimental Setup

These experiments were carried out in the scaled engine cabin test platform (Figure 1). The top and bottom plates of cabin were 8.5 m (L) × 7.5 m (W) and 8.5 m (L) × 6 m (W). The height of cabin was 1.9 m. Inside the cabin there were two gas turbine models with a height of 1.25 m. The experimental bench smoke exhaust system was based on the original ventilation system, based on the principle of smoke control, and improved by adding pipelines and valves. There were four air supply inlets and two smoke exhaust outlets in the engine room. The schematic diagram of the air supply inlet was shown in Figure 1, its size could be adjusted, and the opening of the air inlet was controlled by the valve (0.2 m × 0.2 m, 0.15 m × 0.15 m, 0.1 m × 0.1 m). The air inlet could be set to three different heights—0.1 m, 0.4 m, 0.7 m—from the bottom plate. The smoke exhaust outlet was located at the top of the cabin, and its size was 0.4 m × 0.4 m.

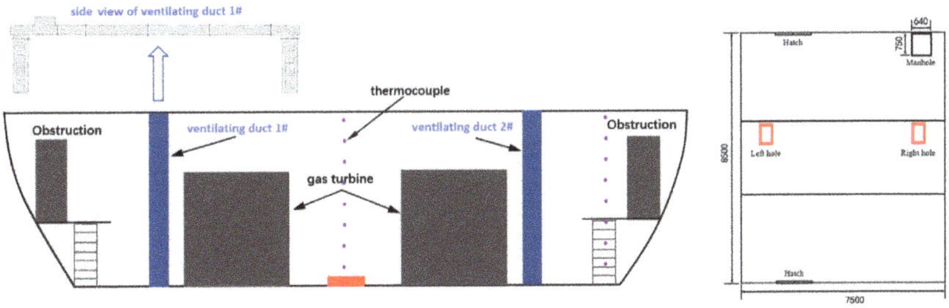

Figure 1. Front view (**left**) and top view (**right**) of the experimental cabin.

To compare the effect of smoke control in the cabin under different forced air conditions, the experiment considered four different air supply configurations of 2.646 m^3/s, 2.352 m^3/s, 2.058 m^3/s and 1.764 m^3/s, and three different air inlet heights of 0.1 m, 0.4 m and 0.7 m. The smoke exhaust volume was based on the existing ventilation system design and is fixed at 0.9 times the air supply volume.

When comparing the cabin smoke control effects under different forced air volumes, the first principle should be to ensure the consistency of the fire source in each test. Therefore, the heptane oil pool fire with good repeatability was used as the experimental fire source. The diameter of the oil pan is D = 40 cm, and the amount of heptane in a single experiment is 3.5 L. The thermophysical properties of heptane are shown in Table 1.

Table 1. Fuel properties.

Name	Heptane
Molecular formula	C_7H_{16}
Density	684 kg/m^3
Flash point	−4 °C
Boiling point	98.5 °C
Heat of combustion	44.6 MJ/kg
Progressive Mass Burn Rate $\dot{m}''_{\infty}$	0.101 (±0.009)
Effective absorption coefficient $k\beta$	1.1 (±0.3)

During the experiment, the Deant ES60K electronic balance was used to measure the mass loss of the pool fire fuel with a range of 60 kg and an accuracy of 0.5 g. Thermocouples were used as temperature sensors in the experiment, and K-type armored thermocouples with a diameter of one millimeter were used for the thermocouples. A total of five thermocouple trees are arranged. The thermocouples are arranged vertically, one at a distance of 0.2 m downward from the height of 1.8 m at the top of the cabin. The specific layout is shown in Figure 2. Thermocouple tree 1, 2 and 3 are placed at the quarter positions on the longitudinal axis of the cabin to monitor the temperature at the fire source in the cabin and on both sides of the fire source. Thermocouple tree 4 is placed at the position of the manhole to monitor the temperature distribution at the manhole as the only escape route after a fire occurs in the cabin. Thermocouple tree 5 is used to monitor the temperature distribution at the left and right ladder openings after a fire occurs in the cabin. Since the left and right are basically symmetrically distributed, only one bunch of thermocouple trees is provided.

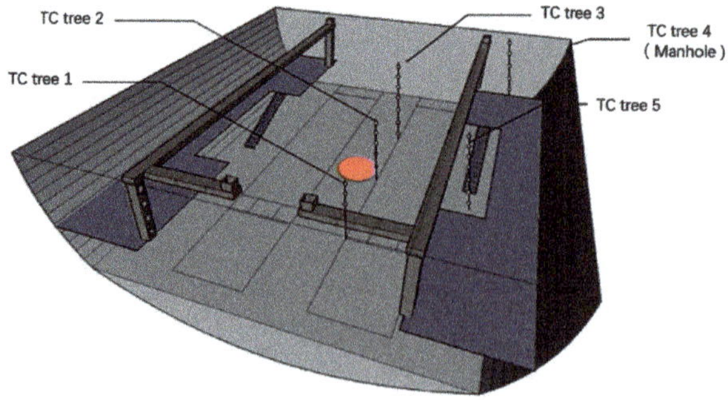

Figure 2. Schematic diagram of fire source and thermocouple layout.

3. Results and Discussion

3.1. Mass Loss Rate

The mass loss rate is a parameter describing the mass loss of fuel during combustion. it can be obtained by differentiating the mass change curve measured by the electronic balance. Through the change in mass loss rate, we can intuitively see the development process of the fire and the influence of boundary conditions on the compartment fire. Figure 3 shows the development law of the mass loss rate of pool fire under the condition of forced air supply and the effect of different air supply air volumes on the mass loss rate. Under the condition of forced ventilation, as the combustion progresses, the temperature of the gas in the cabin rose, and the heat radiation received by the oil pool gradually increased, which accelerated the evaporation process of the fuel and increased the mass loss rate. After burning for a certain time, the pool fire entered the boiling burning stage, and the mass loss rate increased suddenly, and then gradually decreased with the consumption of fuel. With the increase in the air supply volume, the development of pool fire was obviously accelerated, and the burning time was shortened. Figure 4 shows the change in the mass loss rate under the condition of different air supply port heights. With the change in air supply inlet height, mass loss rate change process was basically the same.

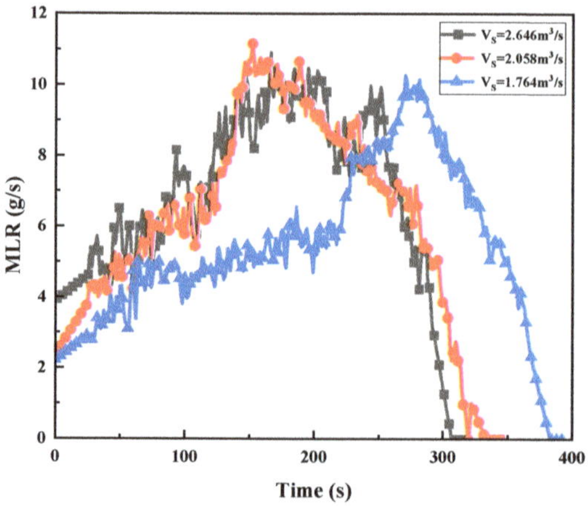

Figure 3. Variation in mass loss rate with time under different air volume conditions.

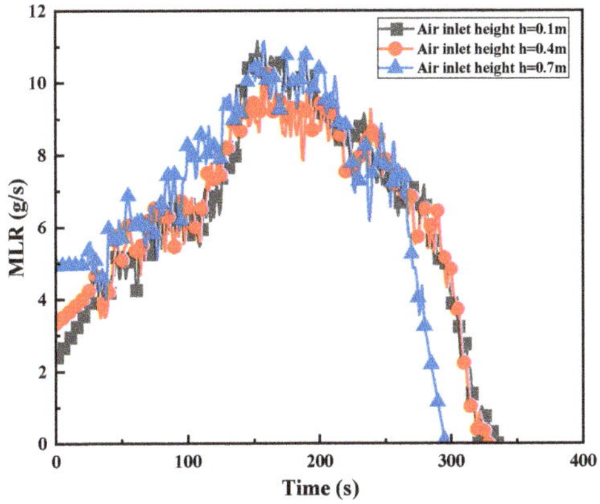

Figure 4. Variation in mass loss rate with time under different air supply inlet heights.

The mass loss rate is time-averaged, and the forced air volume and mass loss rate are dimensionless based on the experimental results in Figure 3 and the experimental results of predecessors [7]. The dimensionless mass loss rate $\dot{m}^*$ can be defined as $\dot{m}^* = \dot{m}/\dot{m}_{th}$, where $\dot{m}$ is experimental mass loss rate, $\dot{m}_{th}$ is the mass loss rate of the oil pool of diameter D in the free atmosphere, $\dot{m}_{th} = \dot{m}''_{\infty}\left(1 - e^{-k\beta D}\right)A_f$. The number of air changes N can be defined as $N = 3600V_s/V_c$, where V_s is the supply air volume, V_c is the cabin volume. By introducing a dimensionless length $f(D, V_c) = D \cdot V_c^{-1/3}$, the relationship between the dimensionless mass loss rate and the number of air changes in this experiment and previous research results can be drawn on a graph, as shown in Figure 5. It can be seen from the figure, under the conditions of the experimental cabin and ventilation configuration, that the fire source mass loss rate and the supplementary air volume have the following relationship:

$$\dot{m}^* f(D, V_c) = -0.21exp\left(-\frac{N}{36 \times 3.25}\right) + 0.24 \tag{1}$$

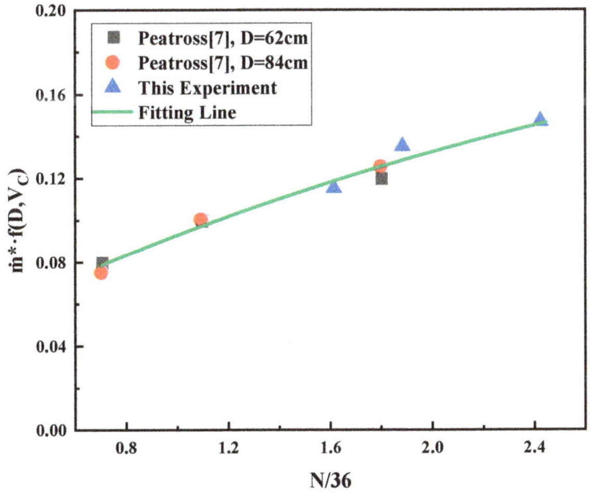

Figure 5. Relationship between dimensionless mass loss rate and dimensionless air volume.

3.2. Temperature Distribution

The heat transfer of the flame to the space gas in the cabin has three ways: conduction, convection and radiation. Among them, the convective heat transfer is the most important, and the convection heat transfer depends on the movement of the smoke and the entrainment formed by the combustion of flame. In the region, the temperature increase is mainly because of smoke. Therefore, analyzing the distribution law of space gas temperature in the combustion process can reflect the distribution characteristics of smoke to a certain extent. Taking the air supply volume of 2.058 m^3/s as an example, Figure 6 shows the temperature changes at different heights in position 1 in the cabin. It can be seen from the figure that the temperature changes at different heights were basically the same. The temperature difference between the measurement points below the height of about 1.2 m in the cabin was not large. Above approximately 1.2 m, the temperature difference between the measurement points increased rapidly. This is because the smoke layer accumulates near the fire source, resulting in a higher temperature in the upper part of the cabin, whereas the temperature of the lower cold air layer does not change much. The cold air layer has a relatively obvious interface.

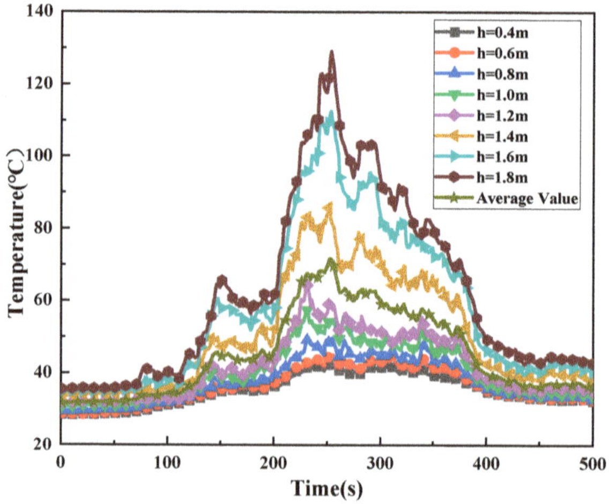

Figure 6. Temperature changes at different heights of the 1# thermocouple tree in the cabin ($V_s = 2.058\ m^3/s$).

Figure 7 shows the variation in vertical temperature with time. At the initial moment, the vertical temperature in the cabin was basically the same. After ignition, with the filling of smoke, the temperature of the upper layer gradually increased, whereas the temperature of the lower layer did not change much. The temperature distribution from top to bottom in the cabin could be regarded as a "double-zone distribution". The temperature inflection points of the upper and lower layers decreased gradually with the settlement of the smoke layer. Because of forced air supply and exhaust, the smoke did not settle to the bottom of the cabin, and the vertical temperature maintained a consistent "double-zone distribution". Afterwards, with the consumption of fuel, the power of the ignition source decreased, the height of the smoke layer gradually increased, and the height of the temperature inflection point of the upper and lower layers gradually increased.

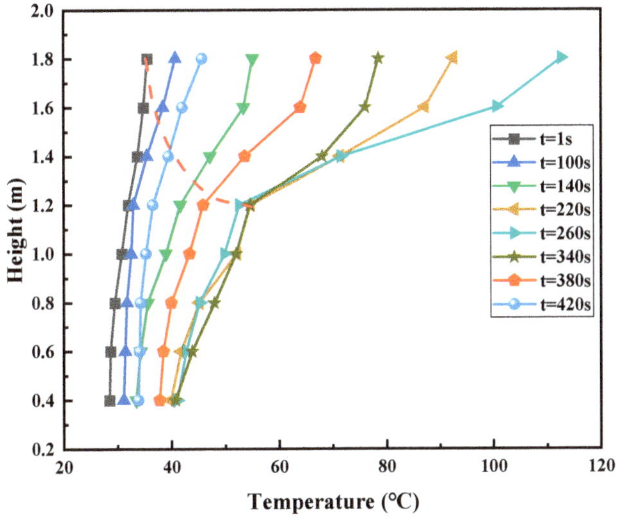

Figure 7. Vertical temperature change in 1# thermocouple tree ($V_s = 2.058\ m^3/s$).

Figure 8 shows the variation in the vertical average temperature at different positions in the cabin. The average temperature here is the average temperature of all thermocouples vertically at this position. It can be seen from the figure that the temperature in the middle part of the cabin was obviously higher than the temperature at the position of the manhole and the entrance of the ladder. Figure 9 shows the variation law of the vertical average temperature at position 1 under different supply air volume conditions. With the increase in air supply air volume, the average temperature in the cabin shown a gradually decreasing trend.

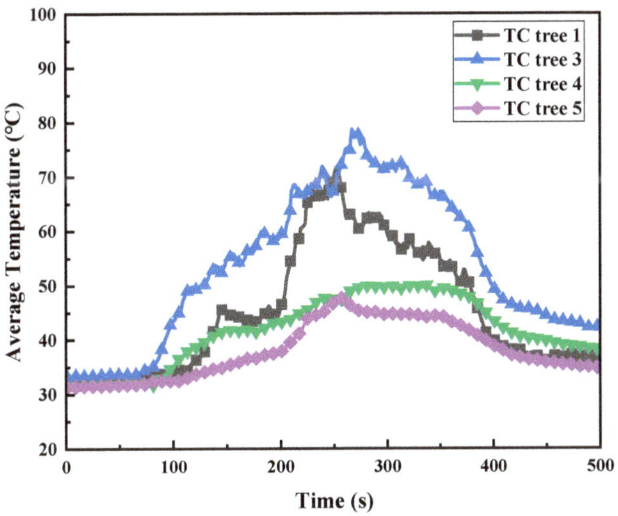

Figure 8. Temperature changes at different positions in the cabin ($V_s = 2.058\ m^3/s$).

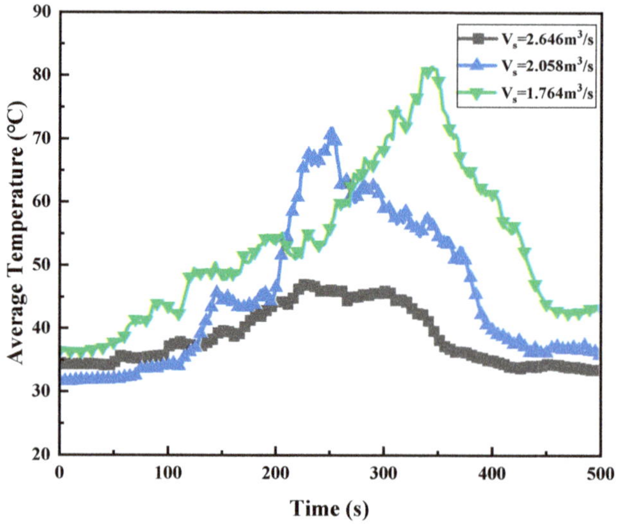

Figure 9. Temperature change at position 1 under different air supply volumes.

3.3. Smoke Height

In fire research, the common methods for solving smoke layer height mainly include: the extinction method, *N* percentage method, minimum integral ratio method and piecewise linear method. Among them, the minimum integral ratio method and the piecewise linear method are used to process the temperature data. In this study, the minimum integral ratio method is used to solve the height of the smoke layer. The method has a relatively clear definition of the neutral plane and a more rigorous calculation and derivation process, and the calculation results are more convincing. It defines the height of the smoke layer as the position where the sum of the minimum integral ratios of the upper and lower layers takes the minimum value. In the calculation, the integral ratios of the upper hot smoke layer and the lower air layer at time t are, respectively:

$$r_u = \frac{1}{(H-z)^2} \cdot \int_z^H T(z)dz \cdot \int_z^H \frac{1}{T(z)} dz \quad (2)$$

$$r_l = \frac{1}{z^2} \cdot \int_0^z T(z)dz \cdot \int_0^z \frac{1}{T(z)} dz \quad (3)$$

Then, the height z of the smoke layer at time t should satisfy:

$$r_t = \min(r_u + r_l) \quad (4)$$

Based on the above method and the temperature data of the 1# thermocouple tree, MATLAB is used to solve the height of the smoke layer, and finally the height of the smoke layer under different air supply configurations is given as shown in Figure 10. It can be seen from the figure that within the range of air volume and air supply port height in this experiment, with the gradual increase in the forced air supply volume, the height of the smoke layer also gradually increases; with the increase in the air supply inlet height, the height of the smoke layer gradually decreases.

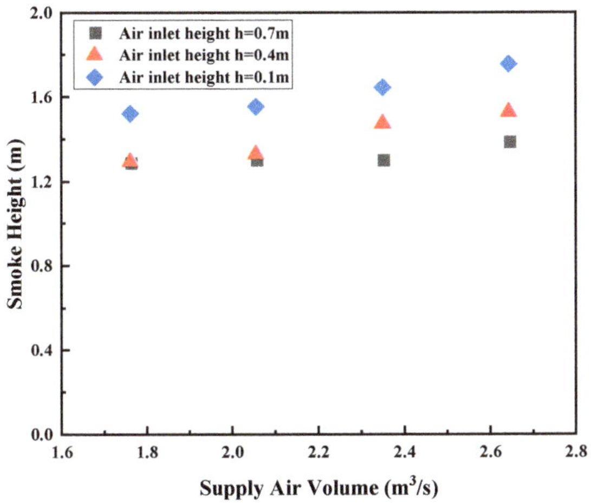

Figure 10. Comparison of cabin smoke layer heights under different supply air volume conditions.

Based on the above data, let the dimensionless smoke layer height $H^* = H/H_0$, the dimensionless air inlet height $h^* = h/H_0$, and introduce the dimensionless parameters P and Q to make the data dimensionless. The curves of P and Q after processing are shown in Figure 11. It can be seen that P and Q are approximately linearly related: $P = 1.72 \times Q + 50.79$.

$$P = F(H^*, N, h^*) = H^*/h^* \times N \times (1.9h^* + 0.7) \tag{5}$$

$$Q = G(N, h^*) = 3.7/h^* \times exp(0.028N) \tag{6}$$

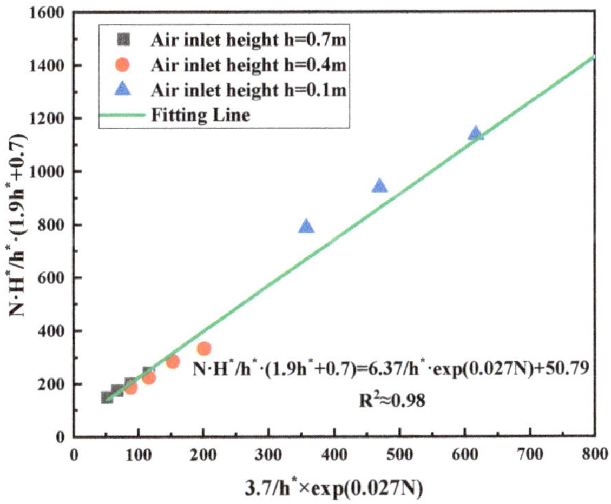

Figure 11. The relationship between the dimensionless smoke layer height and the dimensionless air supply volume.

Based on the above results, the empirical formula for the calculation of the dimensionless smoke layer height under the experimental conditions can be given:

$$H^* = \frac{6.37 \times exp(0.028N) + 50.79h^*}{(1.9h^* + 0.7) \times N} \tag{7}$$

3.4. Prediction of Smoke Layer Height in the Engine Room

In the study of enclosed compartment smoke filling, predecessors usually regarded the ignition source as a point source heat source. Since the object we study is the development of the smoke plume outside the combustion area, we do not consider the combustion process during the fire and the changes in the smoke composition caused by chemical reactions. In actual engine room fire, there is no clear boundary between the smoke layer and the air layer. In the theoretical analysis, the transition layer in the smoke filling process is usually ignored, and only the upper smoke layer and the lower air layer are considered. In addition, this paper assumes that the effect of mechanical air supply on smoke deposition is mainly reflected in the following two aspects: the promotion of mechanical air supply to the fire source; the mechanical air supply promotes the mass mixing of the smoke layer and the lower air layer.

Based on the above assumptions, we choose a physical model with the same size and structure as the experimental cabin. It is assumed that the communication between the interior of the cabin model and the outside world is only through the air supply vents and exhaust vents, and there are no other openings. If the smoke is used as the control body, as shown in Figure 12, the control body always obeys the mass conservation equation during the smoke settling process. There are three main aspects in the process of quality change in the control body: (1) The smoke generated at the fire source enters the control body to increase the quality of the smoke in the control body; (2) the fresh gas provided by the mechanical air supply enters the smoke layer; (3) the upper mechanical exhaust port removes part of the smoke. The main energy transfer process in the cabin can be assumed to be that the heat generated by the fire source in the cabin is divided into three parts: a part of the heat is used to heat the gas in the cabin; a part of the heat is dissipated outward through the wall; a part of the heat is dissipated outward through the smoke outlet.

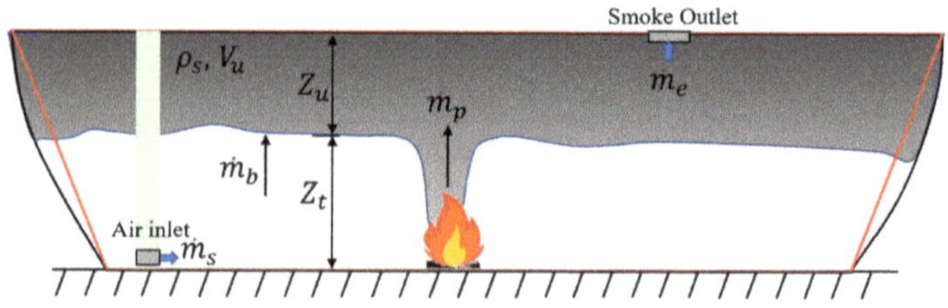

Figure 12. Schematic diagram of the smoke stratification in the engine cabin.

Therefore, its conservation equation can be expressed as:

$$\dot{m}_p + \dot{m}_b - \dot{m}_e = \frac{d\rho_s V_u}{dt} \tag{8}$$

where:

$\dot{m}_p$: the mass flow rate of the smoke entering the upper control body from the fire source area;
$\dot{m}_b$: the mass flow rate of the air entering the smoke layer from other areas;
$\dot{m}_e$: the mass flow rate of the smoke discharged by the mechanical exhaust;
ρ_s: the smoke density, which is assumed to be constant and has the same value with air;
V_u: the volume of the smoke layer.

The volume change rate dV_u of the control body is the volume of the micro-element above the control body. Since the thickness dz of the micro-element is infinitely small, the volume of the entire micro-element can be calculated according to the volume of the cube, that is, $dV_u = A_z dZ_u$, where Z_u is the smoke layer thickness, A_z is the cross-sectional area of the cabin at the lower boundary of the smoke layer.

The Formula (9) can be rewritten as:

$$\frac{dZ_u}{dt} = \frac{\dot{m}_p + \dot{m}_b - \dot{m}_e}{A_z \rho_s} \tag{9}$$

During the combustion process, the hot smoke entrains the surrounding air during the upward movement to form a plume. In the plume theory, there is a law of conservation of mass between the amount of smoke and the entrained air, so the amount of smoke generated depends on the amount of entrained air. In this model, Zukoski's axisymmetric plume model is mainly used, as shown in the following formula.

$$\dot{m}_p = 0.21\left(\frac{\rho_0 g}{C_P T_0}\right)^{1/3} \dot{Q}_C^{1/3} Z^{5/3} = C\dot{Q}_C^{1/3} Z^{5/3} \tag{10}$$

where:

ρ_0: air density;
g: gravitational acceleration;
C_P: specific heat capacity at constant pressure;
T_0: the environment temperature;
$\dot{Q}_c$: the convective part of ignition power;
Z: The height of the plume away from the fire surface.

Based on previous studies, mechanical ventilation has a promoting effect on the power of the fire source. Therefore, combined with Equation (1), we can assume that the mass loss rate $\dot{m}$ of the fire source can be written as:

$$\dot{m} = f(V_s) \tag{11}$$

Therefore, the convection part $\dot{Q}_c$ of the fire power can be written as:

$$\dot{Q}_c \approx 0.7\dot{Q} = 0.7\chi \cdot f(V_s) \cdot \Delta H_c \tag{12}$$

where:

$\dot{Q}$: total heat release rate of the fire source;
χ: combustion efficiency;
ΔH_c: heat of combustion.

In addition, the fresh gas provided by the mechanical ventilation is not only entrained by the fire source, but the gas in other areas will be mixed with the upper layer smoke due to the gas flow. It is assumed that the proportion of the gas mixed with the upper layer smoke in the mechanical supply air volume is ψ. For different air supply heights, ψ is different; thus, it can be assumed that $\psi = g(h)$, then:

$$\dot{m}_b = \psi \dot{m}_s = g(h)\dot{m}_s = g(h)\rho_s V_s \tag{13}$$

During the process of smoke filling, the actual cross-sectional area of the cabin is constantly changing. To give a suitable engineering calculation model, it is assumed that the cross-sectional area of the cabin decreases linearly from top to bottom, as shown in the above figure, so the cross-sectional area of the cabin can be written as:

$$A_z = kz + b \tag{14}$$

The values of k and b in the formula are determined by the size of the cabin. By substituting Equations (11), (13)–(15) into Equation (10), and introduce the height Z_t of the smoke layer, we have:

$$-\frac{dZ_t}{dt} = \frac{dZ_u}{dt} = \frac{C(0.7\chi \cdot f(V_s) \cdot \Delta H_c)^{1/3} Z_t{}^{5/3} + g(h)\rho_s V_s - \dot{m}_e}{(kZ_t + b)\rho_s} \quad (15)$$

Therefore, for the prediction of the height of the smoke layer at different air supply air volumes and air supply port heights, there are:

$$\begin{cases} -\frac{dZ_t}{dt} = \frac{dZ_u}{dt} = \frac{C(0.7\chi \cdot f(V_s) \cdot \Delta H_c)^{1/3} Z_t{}^{5/3} + g(h)\rho_s V_s - \dot{m}_e}{(kZ_t + b)\rho_s} \\ when\ t = 0\ ,\ Z_t = H \end{cases} \quad (16)$$

Based on the results of the experiment and the above theoretical analysis, the results of the smoke layer height prediction are given below. Based on the results obtained from the dimensionless mass loss rate, namely, Equations (1) and (2), the fire source power can be written as:

$$\dot{Q}_c \approx 0.7\chi \cdot \dot{m} \cdot \Delta H_c = 0.7\chi \cdot \left(-204.54 \cdot e^{-V_s/0.37} + 7.11\right) \cdot \Delta H_c \quad (17)$$

The above formula can be applied to the fire source power under different air volume conditions to determine the height of the smoke layer. Formula (17) can be written as:

$$\begin{cases} \frac{dZ_t}{dt} = -\frac{C\left[0.7\chi \cdot \left(-204.54 \cdot e^{-\frac{V_s}{0.37}} + 7.11\right) \cdot \Delta H_c\right]^{1/3} z^{\frac{5}{3}} + \psi V_s \rho_a - V_e \rho_s}{(kz + b)\rho_s} \\ when\ t = 0\ , Z_t = 1.9 \end{cases} \quad (18)$$

where:

C: according to Zukoski's model can be calculated, $C = 0.076432$;
k: k depends on the compartment structure, $k = 6.7$;
b: b depends on the compartment structure, $b = 51$;
ψ: the mixing rate, a function of the air supply port height h, $g(h) = -0.25e^{-\frac{h}{0.64}} + 0.76$.

Put different air volume and air inlet height into the above formula, and iteratively solve based on the second-order Runge–Kutta method. The comparison results between the experimental value and the theoretical value are shown in Figure 13. Compared with the experimental results, it is found that within the range of the experimental air volume, the theoretical prediction results of smoke layer height are consistent with the experimental results, and the relative error is less than 11%. It can be considered that the theoretical prediction model given in this paper has a good prediction effect.

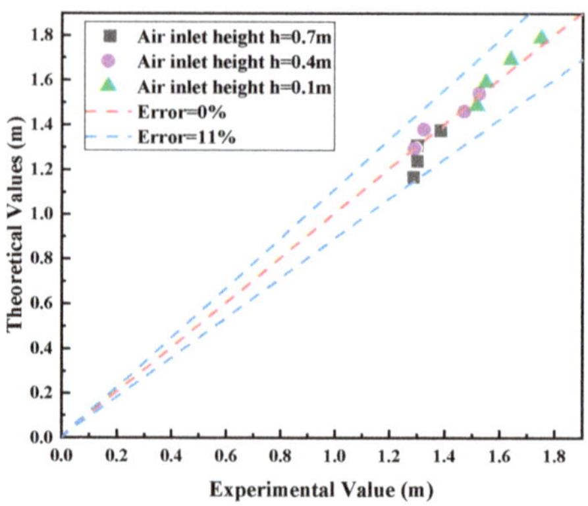

Figure 13. Comparison of predicted and experimental values of smoke layer height under different supply air volume conditions.

4. Conclusions

In this paper, experimental and theoretical research were carried out on the height of smoke layers from engine room fires under the condition of forced ventilation. The development of the cabin room fire and the process of smoke settlement under the conditions of different air supply volume and air supply inlet heights were studied by experiment method. The theoretical analysis was carried out to predict the height of the cabin fire smoke layer under different air supply conditions and verified the validity of the model.

The main conclusions are as follows:

(1) The fire experiment was carried out in the small-scale engine room. Experimental results show that the mass loss rate of the fire source increased approximately exponentially with the increase in the supply air volume. Combined with the experimental data and previous data, the empirical formulas of mass loss rate and air supply volume are given, and they are applied to the smoke layer height prediction model proposed in this study.

(2) The smoke layer height variation curve under different air supply conditions was given based on the minimum integral ratio method. Under the experimental conditions, the height of the smoke layer gradually increases with the increase in air supply volume and the height of the smoke layer gradually decreases with the increase in the height of the air supply inlet.

(3) A prediction model of smoke layer height under different forced ventilation conditions was constructed through theoretical analysis based on the conservation equations. Within the range of the experimental air supply volume and the height of the air inlet, the relative error between theoretical prediction results and experimental results was less than 11%, which could effectively predict the smoke layer height in the in-ship cabin fire. The research is of great significance to the research on the fire risk of the engine room, and provides a theoretical basis for the prediction of the fire risk of the engine room.

Author Contributions: Methodology, X.W.; software, W.Y.; validation, Y.Z.; formal analysis, X.W.; investigation, J.J.; resources, X.W.; data curation, Y.Z.; writing—original draft preparation, X.W.; writing—review and editing, X.C. and S.L. All authors have read and agreed to the published version of the manuscript.

Funding: The National Natural Science Foundation of China (Grant No.51704268) and Fundamental Research Funds for the Central Universities (No. WK2320000048).

Institutional Review Board Statement: Not applicable.

Informed Consent Statement: Not applicable.

Data Availability Statement: Not applicable.

Acknowledgments: This work was supported by the National Natural Science Foundation of China (Grant No.51704268) and the Fundamental Research Funds for the Central Universities (No. WK2320000048).

Conflicts of Interest: The authors declare no conflict of interest.

References

1. El-Helw, M.; Fayed, M.; El-Shobaky, A. Studying different scenarios of operating air conditioning system in smoke management using computational fluid dynamics in naval ships. *Therm. Sci.* **2018**, *22*, 2973–2986. [CrossRef]
2. Zhang, J.; Zhang, B.; Ji, K.; Yan, B.; Fan, M.; Li, W.; Li, Q. Numerical Study on Fire characteristics in Force-Ventilated Compartment with different air inlet locations. In Proceedings of the Journal of Physics: Conference Series, Taiyuan, China, 21–22 October 2017; p. 012028. [CrossRef]
3. Wu, Z.; Wang, L.; Su, S.; Liu, W. Experimental and simulation study of the interaction characteristics of nano-MgO and smoke in the fire of a ship engine room. *Fire Mater.* **2022**, *46*, 953–967. [CrossRef]
4. Zhang, B.; Zhang, J.; Wang, X.; Lu, S.; Li, C.; Chen, R. Effects of air inlet configuration on forced-ventilation enclosure fires on a naval ship. *Fire Technol.* **2016**, *52*, 547–562. [CrossRef]
5. Alvares, N.; Foote, K.; Pagni, P. Forced ventilated enclosure fires. *Combust. Sci. Technol.* **1984**, *39*, 55–81. [CrossRef]
6. Chow, W.K.; Chan, W.L. Experimental studies on forced-ventilated fires. *Fire Sci. Technol.* **1993**, *13*, 71–87. [CrossRef]
7. Peatross, M.J.; Beyler, C.L. Ventilation effects on compartment fire characterization. *Fire Saf. Sci.* **1997**, *5*, 403–414. [CrossRef]

8. Le Saux, W.; Pretrel, H.; Lucchesi, C.; Guillou, P. Experimental study of the fire mass loss rate in confined and mechanically ventilated multi-room scenarios. *Fire Saf. Sci.* **2008**, *9*, 943–954. [CrossRef]
9. Mitler, H.E. Zone Modelinq of Forced Ventilation Fires. *Combust. Sci. Technol.* **1984**, *39*, 83–106. [CrossRef]
10. Foote, K.; Pagni, P.; Alvares, N. Temperature correlations for forced-ventilated compartment fires. *Fire Saf. Sci.* **1986**, *1*, 139–148. [CrossRef]
11. Deal, S.; Beylert, C. Correlating preflashover room fire temperatures. *J. Fire Prot. Eng.* **1990**, *2*, 33–48. [CrossRef]
12. Chow, W.K. Modelling of forced-ventilation fires. *Math. Comput. Model.* **1993**, *18*, 63–66. [CrossRef]
13. Brohez, S.; Caravita, I. Fire induced pressure in airthigh houses: Experiments and FDS validation. *Fire Saf. J.* **2020**, *114*, 103008. [CrossRef]
14. Li, J.; Beji, T.; Brohez, S.; Merci, B. CFD study of fire-induced pressure variation in a mechanically-ventilated air-tight compartment. *Fire Saf. J.* **2020**, *115*, 103012. [CrossRef]
15. Li, J.; Prétrel, H.; Suard, S.; Beji, T.; Merci, B. Experimental study on the effect of mechanical ventilation conditions and fire dynamics on the pressure evolution in an air-tight compartment. *Fire Saf. J.* **2021**, *125*, 103426. [CrossRef]
16. Backovsky, J.; Foote, K.L.; Alvares, N.J. *Temperature Profiles in Forced-Ventilation Enclosure Fires*; Lawrence Livermore National Laboratory: Livermore, CA, USA, 1988.
17. Hayashi, Y.; Hasemi, Y.; Ptchelintsev, A. Smoke Layer Formation by Fires In Forced Ventilation Enclosure. *Fire Saf. Sci.* **2000**, *6*, 805–816. [CrossRef]
18. Zukoski, E.E. Development of a stratified ceiling layer in the early stages of a closed-room fire. *Fire Mater.* **1978**, *2*, 54–62. [CrossRef]
19. Haouari-Harrak, S.; Mehaddi, R.; Boulet, P.; Koutaiba, E. Evaluation of the room smoke filling time for fire plumes: Influence of the room geometry. *Fire Mater.* **2020**, *44*, 793–803. [CrossRef]
20. Zhang, Y.; Liu, Z.; Lin, Y.; Fu, M.; Chen, Y. New approaches to determine the interface height of fire smoke based on a three-layer zone model and its verification in large confined space. *Fire Mater.* **2020**, *44*, 130–138. [CrossRef]
21. National Fire Protection Association. *NFPA 92B. Guide for Smoke Management Systems in Atria, Covered Malls, and Large Areas*; National Fire Protection Association: Quincy, MA, USA, 2005.
22. Tanaka, T.; Yamana, T. Smoke control in large scale spaces (Part 1: Analytic theories for simple smoke control problems). *Fire Sci. Technol.* **1985**, *5*, 31–40. [CrossRef]
23. Zukoski, E.E.; Kubota, T.; Cetegen, B. Entrainment in fire plumes. *Fire Saf. J.* **1981**, *3*, 107–121. [CrossRef]
24. Li, Y. Smoke Flow and control in Large Space Atrium Buildings. Ph.D. Thesis, University of Science and Technology of China, Hefei, China, 2001.
25. Yuan-Zhou, L.I.; Liu, Y. Study on smoke filling process in shipengine room based on full-scale tests. *J. Saf. Environ.* **2018**, *18*, 542–547.
26. Wang, J.; Zhang, R.; Wang, Y.; Shi, L.; Zhang, S.; Li, C.; Zhang, Y.; Zhang, Q. Smoke filling and entrainment behaviors of fire in a sealed ship engine room. *Ocean Eng.* **2022**, *245*, 110521. [CrossRef]

fire

MDPI

Article

Effect of Wind Speed on the Natural Ventilation and Smoke Exhaust Performance of an Optimized Unpowered Ventilator

Mao Li [1], Yukai Qiang [2,*], Xiaofei Wang [3], Weidong Shi [2], Yang Zhou [2] and Liang Yi [2,*]

1 Headquarters of Mazhao Highway Project, Kunming 650034, China; limao9025@hotmail.com
2 School of Civil Engineering, Central South University, Changsha 410075, China; shiweidong@csu.edu.cn (W.S.); zyzhou@csu.edu.cn (Y.Z.)
3 Anhui Transport Consulting & Design Institute Co., Ltd., Hefei 230088, China; wangxiaofei@acdi.ah.cn
* Correspondence: yukaiqiang@csu.edu.cn (Y.Q.); yiliang@csu.edu.cn (L.Y.)

Abstract: Natural ventilators can maintain the ventilation of buildings and tunnels, and can exhaust fire smoke without requiring energy. In this study, we optimized a natural ventilator by adding axial fan blades (equivalent to adding a fan system) to investigate the effect of wind speed on the ventilation and smoke exhaust performance of an optimized natural ventilator. The experimental results showed that the best configuration of the ventilator was five fan blades at an angle of 25° with set-forward curved fan blades. With this configuration, the ventilation volume of the optimized natural ventilator was increased by 11.1%, and the energy consumption was reduced by 2.952 J. The third experiment showed that, in the case of a fire, the optimized ventilator can reduce the temperature of the ventilator faster than the original ventilator, indicating better smoke exhaust performance. The reason for this effect is that, when the optimized natural ventilator rotates, the rotation of the blades creates a flow field with a more evenly distributed wind speed. The experiments proved that natural ventilators can be optimized by adding a fan system. The results of this study can be applied to effectively improve the ventilation performance of natural ventilators to quickly exhaust smoke in building and tunnel fires, and provide a reference for related research on natural ventilators.

Keywords: optimized unpowered ventilator; axial fan blade; wind speed; ventilation performance; energy consumption; smoke exhaust performance

Citation: Li, M.; Qiang, Y.; Wang, X.; Shi, W.; Zhou, Y.; Yi, L. Effect of Wind Speed on the Natural Ventilation and Smoke Exhaust Performance of an Optimized Unpowered Ventilator. *Fire* **2022**, 5, 18. https://doi.org/10.3390/fire5010018

Academic Editor: Dahai Qi

Received: 20 December 2021
Accepted: 22 January 2022
Published: 28 January 2022

Publisher's Note: MDPI stays neutral with regard to jurisdictional claims in published maps and institutional affiliations.

1. Introduction

Ventilators are a type of equipment used to promote air flow [1]. They are classified into two categories: power and natural ventilators [2]. Natural ventilators, also known as unpowered ventilators, are equipment that use the pressure difference between the inside and outside of a building to drive the air flow inside the building [3,4]. Natural ventilators can considerably reduce building energy consumption [5]. Therefore, investigations into optimizing natural ventilators are necessary [6,7]. Gonzalez et al. studied the ventilation performance of a natural ventilator under free flow conditions, and established its mathematical model [8]. Kang et al. determined the relationship of air velocity with ventilators [9]. Kim et al. found that the speed and ventilation volume of natural ventilators are positively correlated with the ambient wind speed [10]. Research results have proven that the optimization of ventilators can improve air quality within a room [11,12].

Studies on the performance optimization of ventilators are important for building ventilation, which were implemented earlier in Europe and America [13–16]. Building openings and various ventilation equipment were the main focus of studies on natural building ventilation [17], and ventilators can effectively improve indoor air quality [18]. Santamouris [19] investigated the relationship of air flow speed and indoor carbon dioxide concentration in classrooms with intermittent natural ventilation. Gan simulated the numerical of the indoor environment [20]. These research results have encouraged further study of air flow, which were also the basis of our study [21,22].

At present, most studies on the ventilation performance of natural ventilators have been conducted through CFD simulation [23–25]. Varela-Boydo and Moya, using CFD, test 28 different inlet extension designs [26]. The results showed that the majority of designs increase air induction to the building by reducing the ascending currents from inside the inlet opening to the outside of the towers to a minimum. However, some of the tested designs formed vortexes inside the inlet openings, which prevented air induction. These results enrich the ventilation theory of natural ventilators, and provide methods to improve the volume of ventilators [27].

However, few scholars have studied the combination of ventilation performance and fan blades. As such, in this study, we investigated the effects of combinations of natural ventilator and axial fan blades on ventilation performance. The design of the combination is shown in Figure 1.

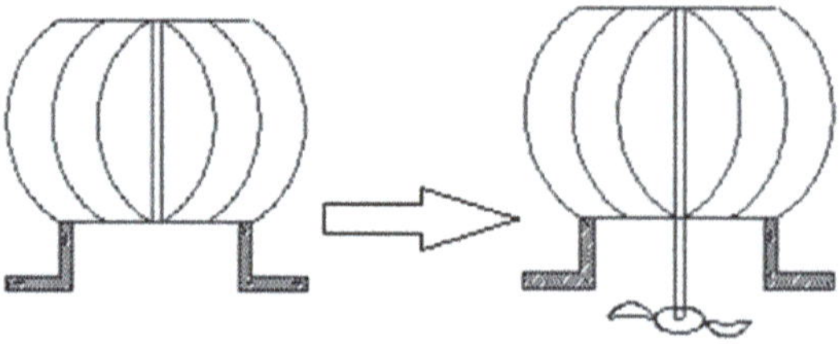

Figure 1. Natural ventilator before and after optimization.

Since the 1950s, scholars have researched fan blades [28]. Beiler found that forward fan blades can improve the ventilation performance of a fan [29]. Hossain optimized the design of aircraft fans [29]. Farrall and Simmons simulated and analyzed the fan of an aircraft [30]. Elgowainy et al. found that the number of fan blades has an important impact on the static pressure of a fan [31]. Yang et al. found that forward fan blades can reduce the pressure loss of axial flow fans, improving fan efficiency [32]. Moreover, Vad et al. found that forward-type fan blades can significantly improve the flow field distribution at the tip of the fan blade, reducing energy loss, as shown in Figures 2 and 3 [33]. The influences of fan blade parameters, such as the number of fan blades, inclination angle, and bending direction, on fan ventilation performance have been investigated.

Figure 2. Photo of forward curved fan blades.

Figure 3. Photo of backward curved fan blades.

In our newly designed natural ventilator (swirl natural ventilator), we used different types of fan blades. We compared and analyzed the influence of design parameters, such as the number of axial fan blades, inclination angle, and bending direction, on ventilation volume and ventilation efficiency. Many research results have been published regarding the optimization of natural ventilators by improving its ventilation performance [34,35], but studies on the use of optimized natural ventilators for smoke exhaust in the case of fire are limited [36]. Therefore, after the ventilation performance of this ventilator was improved through experiments, the proposed ventilator can be installed in buildings or tunnels. As such, when a fire occurs, the ventilator can quickly exhaust the smoke and reduce the temperature in the environment as much as possible, without consuming energy.

Our ventilator design considerably improves the ventilation volume of natural ventilators, thus enhancing the smoke exhaust performance in building or tunnel fires, and reducing the energy consumption of the building.

2. Materials and Methods

2.1. First Experiment

The first experiment focused on the axial fan blade setting. In this experiment, we used a swirling natural ventilator, as shown in Figure 1.

2.1.1. Experimental Model

The rotating shaft of the ventilator was connected to the motor. The axial fan blades and ventilator were coaxial during operation. In order to control the speed of the ventilator, a small adjustable motor (YVF-6314, Shenyang Weishida, Figure 4) was selected to drive the natural ventilator to rotate. The speed of the motor was controlled by a frequency modulator, as shown in Figure 5.

Figure 4. Photo of YVF-6314 adjustable motor.

Figure 5. Photo of frequency modulator.

The experiment was conducted using a controlled variable method. After the power was turned on, we ensured the ventilator reached the predetermined speed by adjusting the frequency modulator. A hot wire anemometer (AR866, Hong Kong Xima, measurement error of ±3% ± 0.1 d) was calibrated to minimize the influence of external environmental factors.

The speed of the natural ventilator (SW6234C, Guangzhou Suwei, measurement error of ±0.05% ± 1 d) was controlled by a tachometer to ensure the stable running of the natural ventilator.

During the experiments, the heat pressure and wind speed were controlled. This experiment was conducted in an air duct model. The model was composed of a stainless steel pipe that was 1.8 m long and 0.8 m in diameter. To reduce air leakage, we used tin foil to seal the gap between the air duct model and the natural ventilator. Considering the size of the air duct model, we selected a QM-600 swirl natural ventilator as the research object, as shown in Figure 6, and its parameters as shown in Table 1. The bottom diameter of the cyclone ventilator was 600 mm, and the ventilator was fixed on the top of the air duct model by a flange and setscrews.

Figure 6. Photo of QM-600 swirl natural ventilator.

Table 1. Parameters of QM-600 swirl natural ventilator.

Type	Manufacturer	Weight (kg)	Maximum Speed (rpm)	Material
QM-600	Shandong TaiLai	2	400	(340#) Stainless steel

Considering the sizes of the air duct model and the natural ventilator, we selected a group of axial flow fan blades with a radius of 30 cm for the study. To reduce the resistance, we selected an axial flow fan composed of plastic. During the experiment, the air duct was fixed 15 cm away from the bottom of the cyclone ventilator by aluminum discs and screws, as shown in Figure 7.

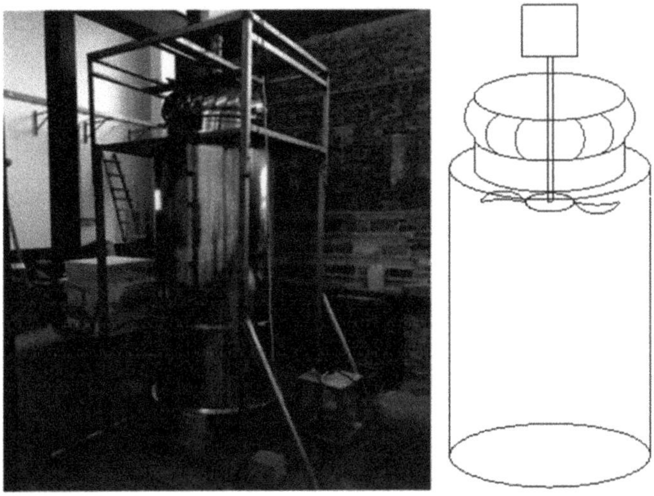

Figure 7. Performance optimization platform of a QM-600 swirl natural ventilator.

2.1.2. Data Measurement

When the fluid flowed in the duct model, the velocity and pressure distribution in the same cross-section of the duct model was uneven due to fluid viscosity. Therefore, the same measurement section of the air duct model was divided into several parts with the same area, and the velocities at characteristic points of each part were measured. The velocities of the characteristic points were used as the average velocity of the part, we used a hot wire anemometer to measure the wind speed.

According to the size of the air duct model, five testing sections were set in the vertical direction in our experiments. The intervals of these measured sections were 0.2, 0.2, 0.4, 0.4, and 0.2 m, and eight testing points were set on each testing section: r_1–r_8.

The air volume in the air duct was calculated according to Equation (1). As the air duct model in this study was a circular tube, we selected the middle rectangle method to arrange the testing points [17].

$$q_v = \sum_{i=1}^{i=n} F_i u_i = \frac{A}{n} \sum_{i=1}^{i=n} u_i \tag{1}$$

where A is the cross-sectional area of air duct model, m^2; F_i is the area of each part, m^2; u_i is the velocity of the ith characteristic point, m/s; and q_v is the average air volume of pipeline, m^3/s.

The cross-section radius of the pipe was assumed to be *r* and the cross-section was divided into *n* concentric rings with equal area, as shown in Figure 8. The radius of the characteristic circles, r_1, r_3, r_5, and r_7, was calculated according to Equation (2), and the radius of the characteristic circles, r_2, r_4, r_6, and r_8, was calculated according to Equation (3).

$$r_{2i-1} = R\sqrt{\frac{2i-1}{2n}} \tag{2}$$

$$r_{2i} = R\sqrt{\frac{2i}{2n}} \tag{3}$$

where *R* is the radius of air duct model, m; *r* is the radius of the characteristic circles in the air duct model, m; *n* is the number of characteristic circles in the air duct model; and *i* is the order of the characteristic circles in the air duct model.

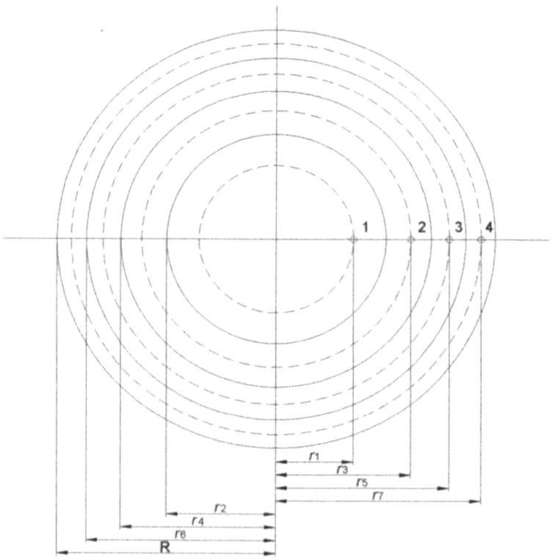

Figure 8. Arrangement of testing points in the air duct model (vertical view).

In the air duct model, we set five testing sections from the top to the bottom. As shown in Figure 9, the distance between testing section 1 and the top was 0.2 m; that between testing sections 1 and 2 was 0.2 m; that between sections 2 and 3, sections 3 and 4, and sections 4 and 5 was 0.4 m; and the distance between section 5 and the bottom was 0.2 m. In each testing section, a total of 8 testing points were arranged. The positions of these testing points were r_8, r_6, r_4, r_2, r_1, r_3, r_5, and r_7 from left to right, as shown in Figure 8. The radii of the characteristic circles were 14.1, 20, 24.5, 28.3, 31.6, 34.6, 37.4, and 40 cm for r_1–r_8, respectively.

Because the position of the natural ventilator was high and the rotating speed of the ventilator was fast, it was difficult to measure the speed of the ventilator. Therefore, we selected a tachometer to measure the ventilator speed. Before the experiment, we pasted a square reflective sticker with a side length of 1 cm on the middle of the outer edge of the ventilator impeller, and we placed the tachometer at an equally high position on the experimental frame with a horizontal distance of 10 cm from the ventilator. In the experiment, when the infrared emission port of the tachometer was facing the center of the natural ventilator, the rotational speed of the natural ventilator could be measured.

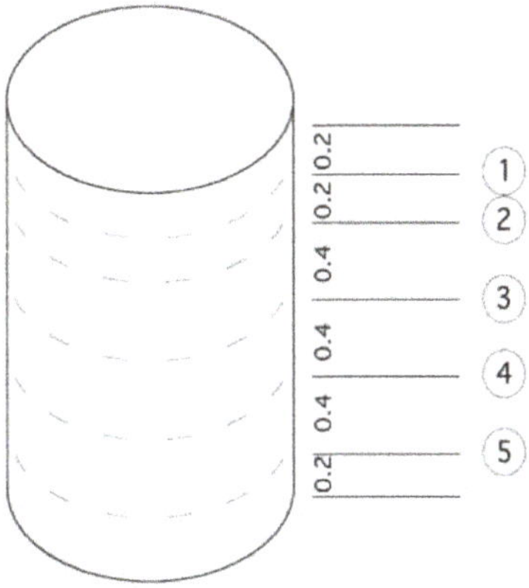

Figure 9. Arrangement of testing points in the air duct model (side view).

2.2. Second Experiment

When the same wind speed was applied to the ventilator (before and after optimization), the ventilator speed changed, so the first experiment could not specifically be used to calculate the change in ventilation efficiency. Therefore, we needed to confirm the relationship between different wind speeds and ventilation speed.

A fan (ZG-2) was used to simulate different wind speed effects. ZG-2 is depicted in Figure 10 and its parameters are listed in Table 2.

Figure 10. Photo of ZG-2 fan.

Table 2. The parameters of ZG fan-2.

Type	Power (KW)	Frequency (Hz)	Air Volume (m^3/h)	Fan Total Pressure (Pa)	Speed (rpm)	Voltage (V)
Zg-2	1.5	50	1000	350	2800	380

2.2.1. Experimental Model

The axial fan was placed to the left of the ventilator, which formed the platform for the second experiment. We used the same natural ventilator speed measurement method in the second experiment as in the first experiment, as shown in Figure 9. The wind speed of the fan was calculated by Equation (5). The experimental setting is shown in Figure 11.

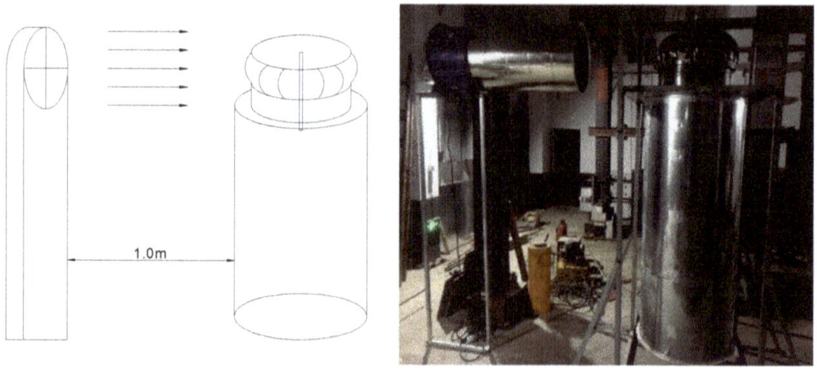

Figure 11. Experimental platform for the fan system.

2.2.2. Data Measurement

In the experiments, we needed to measure wind speed. However, the wind speed was not easy to measure. Because the wind near the outlet of the fan was relatively strong, and the different wind speeds led to an irregular distribution of wind speed within the outlet of the fan, we needed to set a rectifier grid to eliminate eddy currents and to ensure a stable flow. The length of the air duct in the rectification section L1 was 1.0 m. The rectifying plate was located in the middle of the air duct in the rectification section, and the length L2 was 0.2 m.

The rectifier in Figure 12 has a honeycomb structure, and was used on the inlet pipe of blower to balance air flow. Its functions by making the fluid smoothly pass through the fan blade, reducing air flow disturbances to the blade and the pressure loss caused by the rotation of air flow.

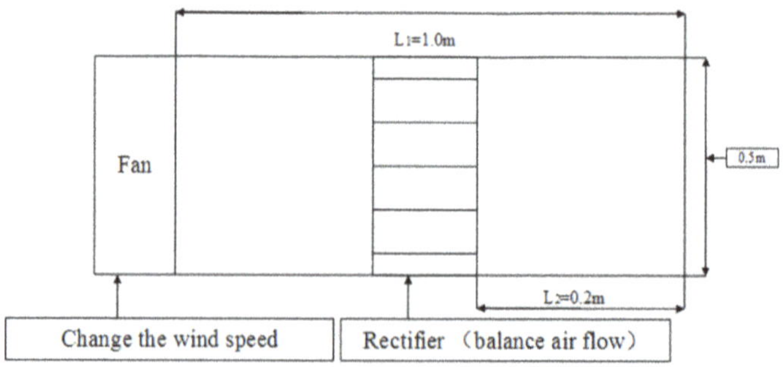

Figure 12. Schematic diagram of the rectifier.

The amount of air flowing out of the outlet in the axial flow fan decayed with increasing air supply distance. Therefore, we calculated the ambient wind speed according to the velocity attenuation equation. The minimum ambient wind speed, which drove the natural ventilator, was determined by several experiments on the amount of air ventilated by the fan. An anemometer was used to measure the wind speed at the fan outlet through the middle rectangle method. Then, the relationship between the ventilation speed and the ambient wind speed could be obtained.

$$\frac{v_x}{v_0} = K\left(\frac{b}{x}\right)^{\frac{1}{2}} \tag{4}$$

where v_x is the wind speed at x distance from the fan, m/s; v_0 is the wind speed at the fan outlet, m/s; K is the proportionality coefficient in the equation, and its value is 2.35; and b is the diameter of the fan, m.

2.3. *Third Experiment*

2.3.1. Experimental Model

The aim of this third experiment was to compare the smoke exhaust performance of the ventilator before and after optimization. We studied smoke exhaust by analyzing the variation in temperature in the air duct of the natural ventilator. During a fire, the faster the temperature decreases in a natural ventilator's air duct, the better its smoke exhaust performance. Therefore, in this experiment, two oil pans (15 × 15 × 4 cm and 20 × 20 × 4 cm) were placed at the bottom of the natural ventilator, and alcohol was poured into it as a fire source. After the alcohol was ignited, the burning metal box simulated a fire. The temperature variation inside the air duct with different fire scales was measured in the experiments. The experimental device is shown in Figure 13.

Figure 13. Photo of fire experiment.

2.3.2. Data Measurement

K-type thermocouples were used for the temperature measurement in the experiments. There were four testing sections in total, and five temperature testing points were symmetrically arranged on each testing section, as shown in Figure 14.

To ensure the accuracy of data measurement, an intelligent temperature acquisition system was used. The measurement error of the system was less than 3%, and the measured data could be stored in real time.

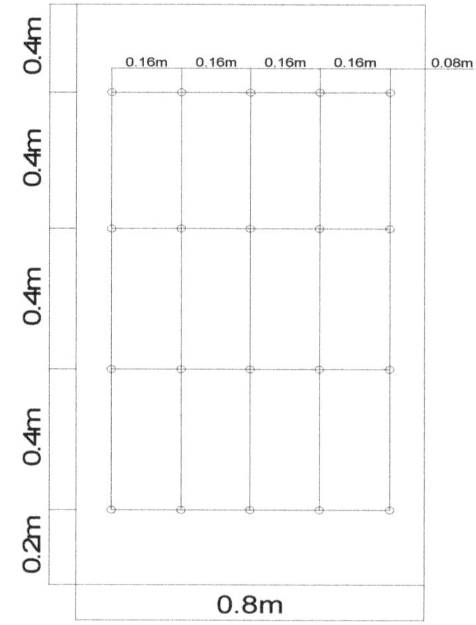

Figure 14. Layout diagram of temperature testing points.

2.4. Measurement Error

The *unbalance rate* refers to the ratio of the difference between the maximum and minimum measured value to the average value of multiple measured values. In this experiment, we used a hot wire anemometer to measure the average wind speed in the air duct. We found that the average wind speed at each testing point fluctuated widely, which was not conducive to experimental measurement. Therefore, to improve and ensure the measurement accuracy of the experimental data, we acquired multiple measurements and checked the *unbalance rate*.

$$Unbalance\ rate = \frac{v_{\max} - v_{\min}}{\overline{v}} \tag{5}$$

$$\overline{v} = \frac{1}{n}\sum_{i=1}^{n} v_i \tag{6}$$

where $v_{\max}$ is the maximum measured value in the air duct, m/s; $v_{\min}$ is the minimum measured value in the air duct, m/s; and v is the average value of the measured data in the air duct, m/s.

3. Results

3.1. Blade Number Optimization

The number of fan blades was set to three, four, five, and six. We compared the performance of the natural ventilator with different number of axial fan blades with that of the original natural ventilator without fan blades. To avoid the influence of other factors, the radius, inclination angle, bending direction, and the position of the fan blades were unchanged. The experimental conditions are listed in Table 3.

Table 3. Experimental setting of optimization of number of fan blades.

Condition Number	Number of Blades	Bending Direction
A01	0	Forward curved
A02	3	
A03	4	
A04	5	
A05	6	

Figure 15a–c shows the ventilation speed and wind speed at different testing points in the air duct.

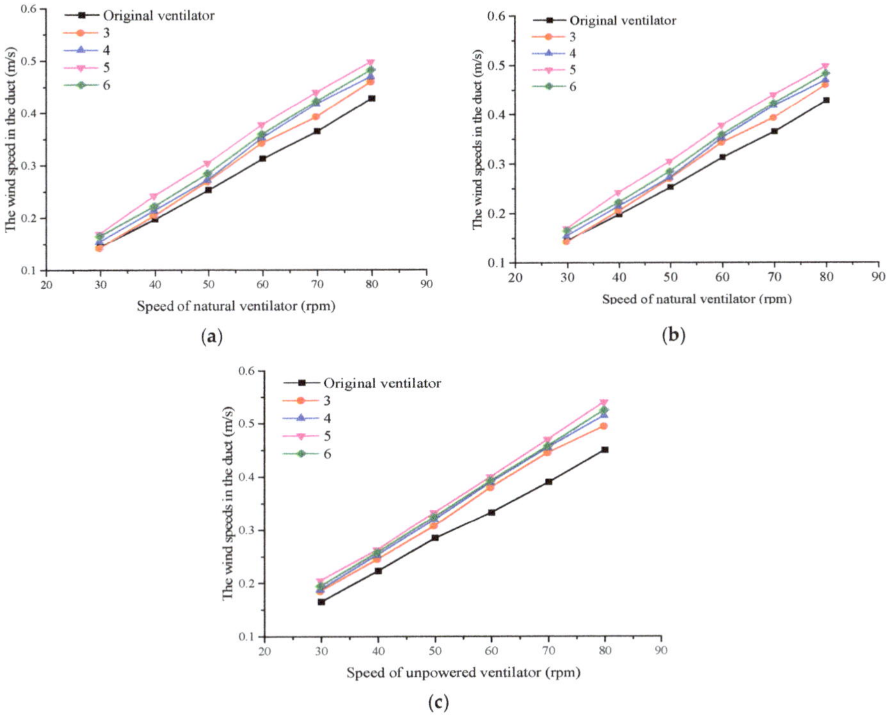

Figure 15. Relationships between wind speed in air duct and natural ventilator speed with different blade numbers at different testing points. (**a**) Testing point 3; (**b**) Testing point 4; (**c**) Testing point 5.

Figure 15a–c shows that the ventilation volume, from high to low, was found for five, six, four, and axial flow fan blades at the same speed. When the number of axial fan blades was increased, the air volume, and thus ventilation performance, generated by the natural ventilator increased first and then decreased. With five axial flow fans, the natural ventilator produced the largest air volume, and its ventilation performance was the best.

According to the principles of fluid mechanics, when a natural ventilator works, the fluid in the duct is pushed by an axial force, which increases the energy of the fluid [37]. When an axial flow fan blade is added, it rotates with the ventilator. In addition, an axial thrust is generated, which increases the ventilation capacity of the whole natural ventilator.

When axial fan blades are used, they should reduce the air collision caused by the rotation of the fan blades as the ventilation rate of the ventilator increases. To ensure the ventilation volume, the fan speed of the ventilator must be increased, which increases energy consumption. After adding three fan blades, the maximum ventilation volume was not obtained in the experiment (as shown in Figure 15c, the maximum wind speed was 0.525 m/s with three blades). However, the addition of five fan blades resulted in the maximum experimental ventilation volume due to the moderate number of blades generating the minimum friction loss in air at a relatively low speed. The air entering the blade could gain more kinetic energy due to the lesser friction, resulting in a higher ventilation rate.

3.2. *Blade Inclination Angle Optimization*

The ventilation performance of the ventilator was the best with five axial flow fans. However, when an axial flow fan works with a natural ventilator, the flow distribution on the rotating plane of the fan blade is not uniform due to the blade inclination angle, which reduces the effect of a ventilator. By comparing the changes in the air volume of the natural ventilator for different fan blade inclination angles, we determined the optimal inclination configuration of the fan blades; the working conditions are shown in Table 4.

Table 4. Experimental setting for the optimization of blade inclination angle.

Condition Number	Angle (Degrees)	Bending Direction
B01	5	Forward curved
B02	15	
B03	20	
B04	25	
B05	30	

The inclination angles of fan blades used in experiments were 5°, 15°, 20°, 25°, and 30°, as shown in Figure 16.

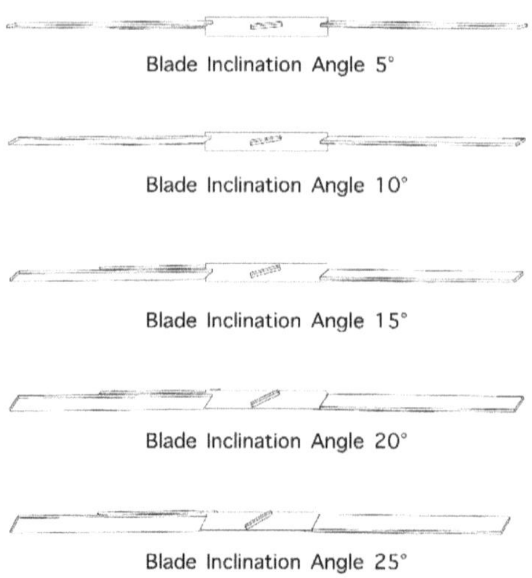

Figure 16. Different blade inclination angles used in the experiment.

Figure 17 shows that when other system parameters of the natural ventilator remained unchanged, the inclination angle of the fan blade and the ventilation rate of the ventilator did not show an linear increasing relationship. When the axial fan blade inclination angle was 25°, the air volume produced by natural ventilator was the largest; the lowest was produced when the inclination angle of fan blade was 5°.

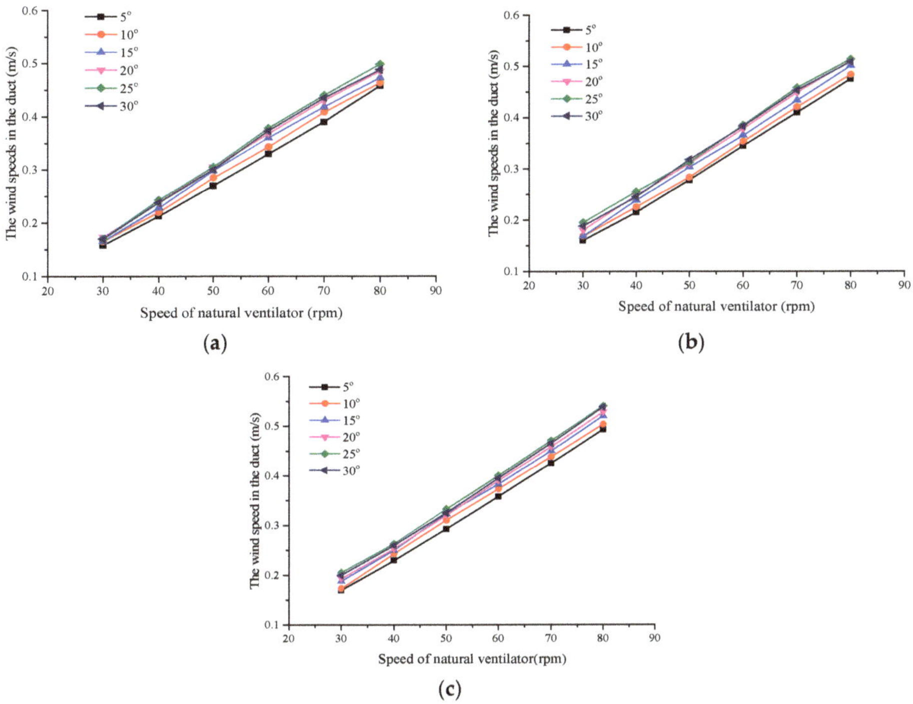

Figure 17. Relationships between wind speed in the air duct and natural ventilator speed with different blade inclination angles at different testing points. (**a**) Testing point 3; (**b**) Testing point 4; (**c**) Testing point 5.

With the increase in the fan blade angle, the pressure difference between the upper and lower surfaces of the fan gradually increased, which enhanced the air disturbance around the fan blade. The ventilation rate was not the maximum when the blade angle was 30° because the pressure between the upper and lower surfaces of the axial flow fan is too large when the fan blade angle is too large. Some friction occurs between the fan blade and the air, and part of the energy is lost. The formation of a flow field on the rotating blade surface is caused by air inflow. When the air collides with the blade surface, part of a secondary vortex may be produced, which reduces the air volume produced by the ventilator and reduces ventilation performance. Therefore, we determined that the optimal blade inclination is 25°.

3.3. Blade Bending Direction Optimization

Combined with the analysis of the previous experimental results, we found that the performance of the ventilator was the best with five axial flow fans and a 25° angle on the fan blades. Then, we studied the influence of the bending direction of the axial fan blade on the ventilation effect. Axial flow blades' bending direction can be divided into forward and backward curved blades. The rotation direction of the forward curved blade is the same as that of the ventilator, whereas that of the backward curved blade is opposite to that of the ventilator. The experimental conditions are shown in Table 5.

Table 5. Experimental setting of optimization of blade bending directions.

Condition Number	Fan Blade Radius (cm)	Bending Direction
C01	30	Forward curved
C02	30	Backward curved

According to Figure 18, we found the flow velocity in the air duct changed with the ventilator speed under different working conditions. Under the same rotating speed, the air volume of the ventilator with forward curved blades was significantly larger than that with the backward curved blade, indicating a better ventilation effect.

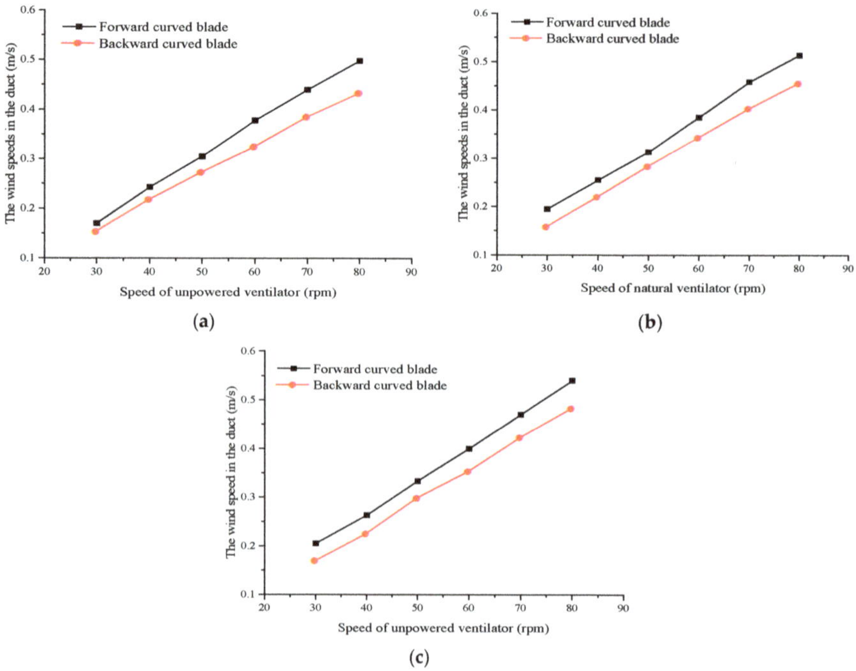

Figure 18. Relationships between wind speed in the air duct and natural ventilator speed with different blade bending directions at different testing points. (**a**) Testing point 3; (**b**) Testing point 4; (**c**) Testing point 5.

When the axial fan blade rotates with a natural ventilator, the forward blade of the axial fan blade contacts the air in the duct. Forward curved blades have a relatively short leading edge within a streamlined fan section, which reduces air drag. The rotating blade strikes the air, with relatively little kinetic energy lost to the fluid. The torsion angle of the back edge of a forward curved fan blade within a wind duct is significantly larger than that of the forward edge, so relatively high wind pressure can be generated. Therefore, this type of fan blade shape reduces the energy loss when a natural ventilator is rotating and improves its ventilation performance.

3.4. Relationship between Wind and Ventilation Speeds

Based on the above experimental analysis, we found that the ventilation effect of the natural ventilator was the best when forward curved axial flow fans are used with a blade inclination of 25° and five blades. Therefore, the experiment conditions were set as described in Table 6.

Table 6. Comparison of fan blade settings between the original and optimized ventilators.

Condition Number	Number of Blades	Number of Blades	Bending Direction
D01	0	-	-
D02	5	25	Forward curved

As shown in Figure 19, we found the speed of the original ventilator increased with increasing ambient wind speed. When the original ventilator rotated with the rated speed of 60 rpm, the wind speed with the fan outlet was 5.18 m/s. However, the same wind speed applied to the optimized ventilator caused rotation at 57.06 rpm. Therefore, compared to the original ventilator, the rotation speed of optimized ventilator was reduced by about 3 rpm.

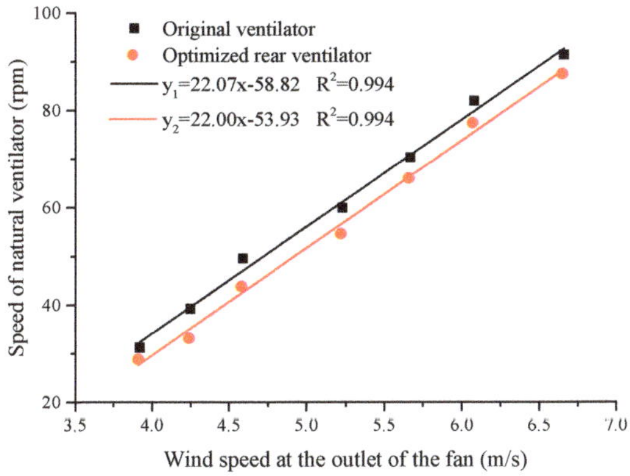

Figure 19. Relationships between wind speed in the air duct and natural ventilator speed before and after optimization.

As mentioned above, when a ventilator works, it can be approximately regarded as a fan system, and the speed of the ventilator is proportional to the environmental wind speed. In the experiment, when fan blades were added, the overall approximation could be regarded as a series superposition of two fan systems. Because the added fan blades are small, the natural ventilator consumes little energy for rotation. When the same wind speed acts on the ventilator, the speed of the ventilator decreases, but the curved slope of ventilator speed with the change in ambient wind speed remains unchanged.

Figure 20 compares the relationships between wind speed in the air duct and natural ventilator speed before and after optimization.

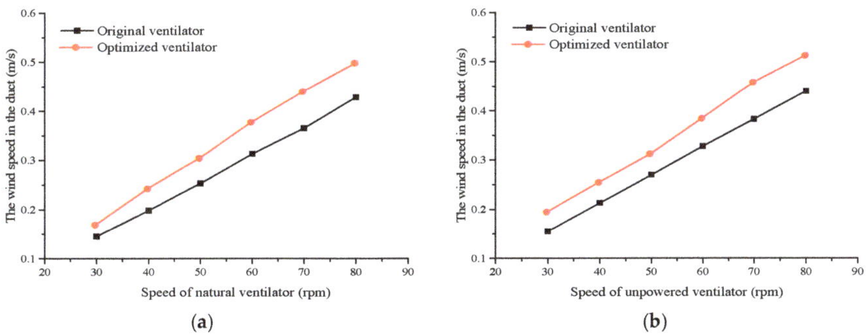

Figure 20. *Cont.*

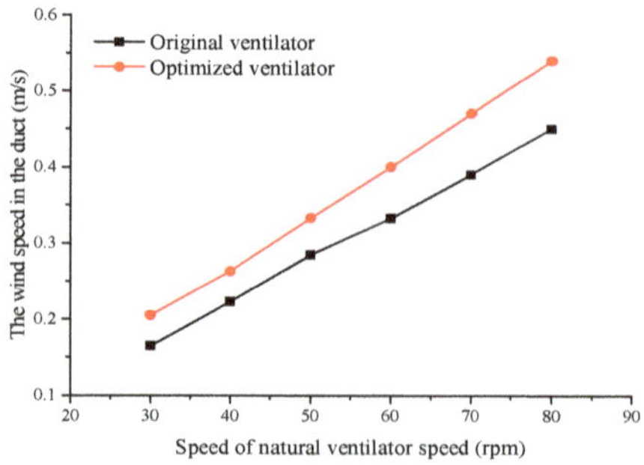

(c)

Figure 20. Relationships between wind speed in the air duct and natural ventilator speed before and after optimization at different testing points. (**a**) Testing point 3; (**b**) Testing point 4; (**c**) Testing point 5.

According to the above experimental results, compared to the original natural ventilator, the ventilation capacity of the newly designed natural ventilator is higher. For example, at point 3, when the original natural ventilator works at 60 rpm, the wind speed into the duct is 0.33 m/s. When the same energy is consumed, the rated speed of the newly designed ventilator is 55.5 rpm and the wind speed in the duct is 0.37 m/s.

3.5. Quantitative Calculation of Optimized Natural Ventilators

3.5.1. Ventilator Energy Savings

When the blades of a natural ventilator rotate, environmental wind on the natural ventilator, which converts the wind energy into the mechanical energy of the ventilator, and the kinetic and potential energy of the fluid. The equation is as follows:

$$W = \Delta E_1 + \Delta E_2 + \Delta E_3 \quad (7)$$

where W is the external environmental energy; ΔE_1 is the variation in kinetic and potential energy before and after fluid flowing through the natural ventilator; ΔE_2 is the mechanical energy consumption of a. natural ventilator during rotation; ΔE_3 is the energy losses, such as fluid flow loss and ventilator rotational friction loss.

$$\Delta E_1 = \frac{1}{2} m_1 \left(v_2^2 - v_1^2\right) + m_1 g (Z_2 - Z_1) \quad (8)$$

where m_1 is the mass of fluid, kg; v_1 and v_2 are the velocity of fluid before and after flowing through natural ventilator, m/s, respectively; Z_1 and Z_2 are the variation in potential energy of fluid before and after flowing through natural ventilator, m, respectively.

Because the air volume loss during the whole process of the fluid flowing through the natural ventilator can be approximately ignored, the temperature change during the whole process of the flow is very small, and the change in fluid density can be ignored, the following flow equation can be obtained:

$$v_2 = v_1 \quad (9)$$

The shape of the natural ventilator was irregular, but it could be approximated as an elliptical cylinder, as shown in Figure 21.

$$\Delta E_2 = \frac{1}{2} I \omega^2 \quad (10)$$

$$I = m_2 r^2 \quad (11)$$

$$r = \frac{L + R}{2} \quad (12)$$

where I is the momentum of inertia about the cylindrical surface, kg·m_2; m_2 is the mass of the natural ventilator, m/s; r is the radius of the cylindrical surface, m; R is the radius of the exhaust outlet at the bottom of the natural ventilator, m; L is the maximum middle radius of the natural ventilator, m; and ω is the angular velocity of the natural ventilator, rad/s.

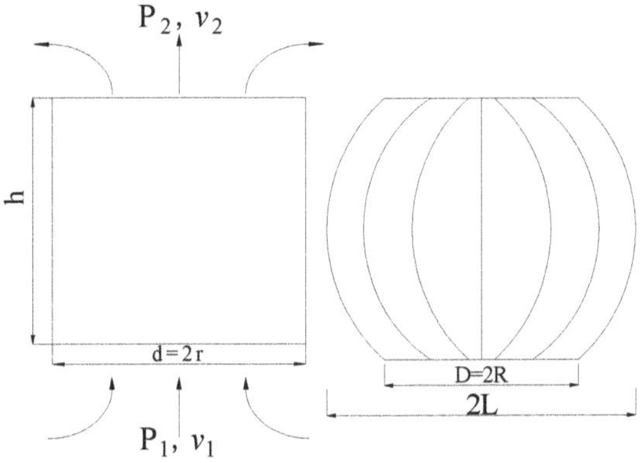

Figure 21. Simplified drawing of a natural ventilator.

For calculation, a single axial fan blade is approximately equivalent to a thin disk with a radius r_i:

$$\Delta E_i = \frac{1}{2} m_i r_i^2 \omega^2 \quad (13)$$

$$r_i = \frac{1}{\sqrt{2}} r_1 \quad (14)$$

where ΔE_i is the rotational kinetic energy of a single axial fan blade, J; m_i is the mass of a single axial fan blade, m/s; r_i is the radius of the thin disk, m; and r_1 is the radius of the axial fan blade, m.

According to the above ventilation calculation equation, the total capacity consumed by the original ventilator was calculated using the parameters in Table 7, as follows:

$$\begin{aligned}
\Delta E_1 &= \tfrac{1}{2} m_1 \left(v_2^2 - v_1^2\right) + m_1 g (Z_2 - Z_1) \\
&= 0 + 0.427\ \text{kg} \times 9.8\ \text{m/s}^2 \times 0.55\ \text{m} = 2.302\ \text{J}\ (v_1 - v_2 = 0) \\
\Delta E_2 &= \tfrac{1}{2} m_2 r^2 \omega^2 \\
&= 1/2 \times 9.6 \times 0.452 \times 6.282 = 38.334\ \text{J}\ (\omega = 2\pi n = 6.28\ \text{rad/s})
\end{aligned}$$

Table 7. Relevant parameters. The parameters of calculating energy savings.

Natural Ventilator	m_2	R	L	q_1	n
	9.6 kg	0.3 m	0.6 m	1200 m^3/h = 0.33 m^3/s	60 r/min = 1 r/s
Fluid	ρ	h	v_1	m_1	
	1.293	$Z_2 - Z_1$ = 0.55 m	Q × t = 0.33 m^3/s × 1 s = 0.33 m^3	$\rho \times v_1$ = 1.293 kg/m^3 × 0.33 m^3 = 0.427 kg	
Axial Fan Blade	m_3	r_1	N	r_i	
	0.1 kg	0.3 m	5	0.212 m	

The total energy (W) that is input by the external environmental system is required when the natural ventilator rotates to the rated speed of 60 rpm:

$$\begin{aligned} W &= \Delta E_1 + \Delta E_2 \\ &= 2.302 + 38.334 = 40.636 \text{ J} \end{aligned}$$

Through experiments, we found the average speed of the optimized natural ventilator n' was 57.06 rpm = 0.951 r/s;

$$\begin{aligned} W &= \Delta E_1 + \Delta E_2 + \Delta E_i = m_{1'} g\,(Z_2 - Z_1) + \tfrac{1}{2} \times m_2 r^2\,(2\pi n')^2 + \tfrac{1}{2} \times m_i r_i^2\,(2\pi n')^2 \times \text{N} \\ &= (0.398n' \times 1.293) \times 9.8 \times 0.55 + \tfrac{1}{2} \times 9.6 \times 0.45^2 \times 4\pi^2 (n')^2 + \tfrac{1}{2} \times 0.1 \times 0.212^2 \times 4\pi^2 \times (n')^2 \times 5 \\ &= 2.578 + 34.705 + 0.401 = 37.684 \text{ J} \end{aligned}$$

According to the above equation, the energy consumption saved by the optimized natural ventilator (ΔE) is

$$\begin{aligned} \Delta E &= \Delta E_1 - \Delta E_2 \\ &= 40.636 - 37.684 = 2.952 \text{ J}. \end{aligned}$$

Therefore, the energy consumption of the optimized natural ventilator was reduced by 2.952 J.

3.5.2. Ventilation Volume

When the fan works, the air volume generated by the fan is proportional to the speed of the fan blades. The ventilation volume calculation equation of the ventilator is as follows:

$$\frac{Q_1}{Q_2} = \frac{n_1}{n_2} = k \tag{15}$$

where Q_1 is the air volume of the original natural ventilator at different speeds, Q_2 is the air volume of the optimized natural ventilator at different speeds, m^3/s; n_1 and n_2 are the fan speeds, rpm; and k is the scale factor.

When the natural ventilator and the axial fan blade rotate, the working principle of the fan system is similar to that of the fan system. The relationship between the air volume and the rotating speed is as follows, and the scale factor k is affected by the shape of the ventilator and the axial fan blade:

$$q = k \cdot n \tag{16}$$

where q is the air flow generated by the rotation of the natural ventilator and axial fan blade, r/s; n is the speed of ventilator, r/s; and k is the scale factor.

According to the analysis of the experimental results, assuming that only the ventilator was working, $q_1 = k_1 \cdot n_1$, we found the scale coefficient k_1 was 0.339. When axial fan blades are added to the ventilator, $q_2 = k_2 \cdot n_2$. We found the scale coefficient k_2 was 0.389. According to the experimental results, k_1 and k_2 are applicable to the different types of fan blade settings.

The original natural ventilator worked at 60 rpm = 1 r/s, and the wind speed in the duct was 0.33 m/s. However, when the same energy was consumed, the average rated speed of the natural ventilator, which had been optimized, was 57.06 rpm = 0.951 r/s, and the wind speed in the duct was 0.37 m/s.

Before the ventilator was optimized, the volume $q_1 = k_1 \cdot n_1$ was $0.339 \times 1 = 0.339$ m^3/s = 1200.4 m^3/h. After the ventilator was optimized, the volume $q_2 = k_2 \cdot n_2$ was $0.389 \times 0.951 = 0.37$ m^3/s = 1333.2 m^3/h. Then, η was calculated as:

$$\eta = \frac{q_2 - q_1}{q_1} = \frac{1333.2 - 1200.4}{1200.4} \times 100\% = 11.1\% \tag{17}$$

The ventilation capacity of the optimized natural ventilator was increased by 11.1%.

3.6. Smoke Exhaust Performance of Optimized Natural Ventilator

The working conditions of fire smoke experiments using the original and optimized ventilator are shown in Table 8.

Table 8. Experimental conditions in the smoke exhaust experiments.

Ventilator	Fuel Quality (kg)	Fuel Thickness (cm)	Oil Pans Size (cm)
Original	1	2	15 × 15 × 4
Optimized	1	2	15 × 15 × 4
Original	1	2	20 × 20 × 4
Optimized	1	2	20 × 20 × 4

According to Figure 22, when the fire was small (15 × 5 × 4 cm), the effects of the optimized and original ventilators on temperature reduction were approximately the same. This is because the thermal pressure driving force was not enough to drive the optimized ventilator when the fire was small. Therefore, the temperature did not show an obvious downward trend.

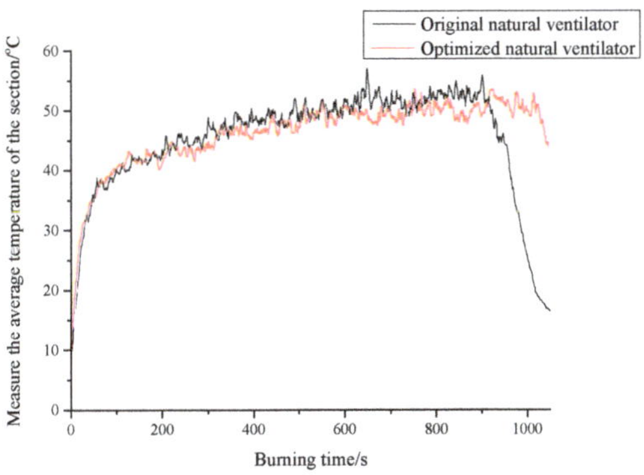

Figure 22. Temperature change of ventilator before and after optimization in the case of a fire environment (fire scale: 15 × 15 × 4 cm).

However, Figure 23 shows that the optimized ventilator markedly reduced the temperature under the thermal pressure driving force of the fire.

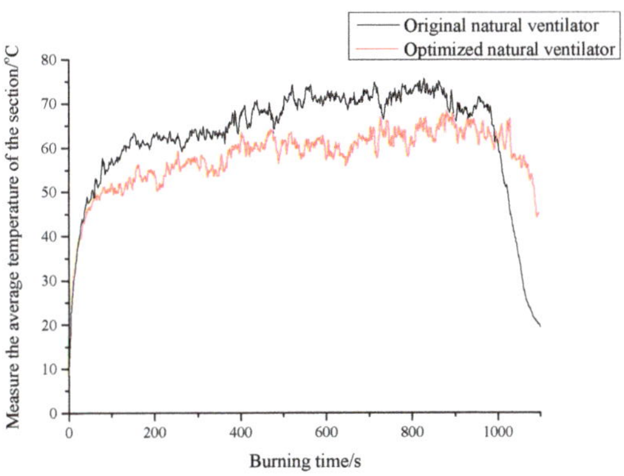

Figure 23. Temperature change of ventilator before and after optimization in fire environment (fire scale: 20 × 20 × 4 cm).

The hot pressure driving force generated by the flame makes the ventilator rotate at a higher speed. Therefore, if the optimized ventilator is used for smoke exhaust, the temperature in the air duct will decrease faster, the temperature generated by the flame can exhaust the smoke faster, and the smoke exhaust efficiency of the ventilator will be improved.

4. Conclusions

To further apply an unpowered ventilator in different buildings or tunnels and quickly exhaust smoke during building and tunnel fires, in this study, we conducted a series of experiments to investigate its ventilation and smoke exhaust performance. Our conclusions are summarized as follows:

(1) After adding a set of axial fan blades below the natural ventilator, the rotational speed of the ventilator decreased at the same power, but the ventilation volume of the ventilator increased, and the ventilation performance of the ventilator was enhanced.
(2) At the same speed, with increasing numbers of axial flow fan blades and increasing fan blade angle, the air volume of the ventilator first increased and then decreased. When the number of fans was five and the angle was 25°, the air volume of the ventilator was the largest and the ventilation effect was the best. Compared to adding backward curved fan blades, the ventilator with forward curved blades produced a larger air volume and a better ventilation effect at the same wind speed.
(3) When the wind speed at the fan outlet reached 5.179 m/s, the original ventilator rotated at the rated speed of 60 rpm. When the same wind speed acted on the optimized ventilator, the speed was 57.06 rpm; the speed was reduced by 4.5 rpm. The air volume was increased by 11.1%, and the energy consumption was reduced by 2.952 J.
(4) The optimized ventilator could quickly exhaust he fire smoke in an actual experiment and lower the temperature of the ventilator without consuming energy. Therefore, the optimized ventilator can be installed in buildings or tunnels to quickly exhaust fire smoke.

The natural ventilator constructed in this study was a swirling natural ventilator 6. This type of unpowered ventilator was also used in the smoke exhaust experiments. Therefore, we conclude that this type of ventilator is more suitable for selected flow natural ventilators, and the optimized natural ventilator is suitable for smoke exhaust from buildings or tunnels. However, in other scenes (such as factories), the optimized unpowered ventilator has not been tested, which is a future research direction for natural ventilator smoke exhaust.

Author Contributions: M.L.: Conceptualization, Methodology; Y.Q.: Formal analysis, Writing-Original Draft; X.W.: Investigation; W.S.: Formal analysis; Y.Z.: Writing-Reviewing and Editing; L.Y.: Supervision, Methodology, Resources. All authors have read and agreed to the published version of the manuscript.

Funding: This research is funded by the Science and Technology Project of Department of Transportation of Yunnan Province under grant No. [2013]261.

Institutional Review Board Statement: Not applicable.

Informed Consent Statement: Not applicable.

Data Availability Statement: Not applicable.

Conflicts of Interest: The authors declare no conflict of interest.

References

1. Linden, P.F. The fluid mechanics of natural ventilation. In Proceedings of the 14th Australasian Fluid Mechanics Conference Adelaide University, Adelaide, Australia, 9–14 December 2001. [CrossRef]
2. Orme, M. Estimates of the energy impact of ventilation and associated financial expenditures. *Energy Build.* **2001**, *33*, 199–205. [CrossRef]

3. Dong, H.-W.; Kim, B.-J.; Yoon, S.-Y.; Jeong, J.-W. Energy benefit of organic Rankine cycle in high-rise apartment building served by centralized liquid desiccant and evaporative cooling-assisted ventilation system. *Sustain. Cities Soc.* **2020**, *60*, 102280. [CrossRef]
4. Evola, G.; Popov, V. Computational analysis of wind driven natural ventilation in buildings. *Energy Build.* **2006**, *38*, 491–501. [CrossRef]
5. Dehghani-sanij, A.R.; Soltani, M.; Raahemifar, K. A new design of wind tower for passive ventilation in buildings to reduce energy consumption in windy regions. *Renew. Sustain. Energy Rev.* **2015**, *42*, 182–195. [CrossRef]
6. Hamid, M. Experimental and numerical study on natural ventilation performance of various multi-opening wind catchers. *Build. Environ.* **2010**, *46*, 370–378.
7. Sha, H.S.; Qi, D.H. A Review of High-Rise Ventilation for Energy Efficiency and Safety. *Sustain. Cities Soc.* **2020**, *54*, 101971. [CrossRef]
8. Gonzalez, M.A. On the aerodynamics of natural ventilators. *Build. Environ.* **1984**, *19*, 179–189. [CrossRef]
9. Kang, J.H.; Lee, S.J. Improvement of natural ventilation in a large factory building using a louver ventilator. *Build. Environ.* **2008**, *43*, 2132–2141. [CrossRef]
10. Kim, Y.S.; Han, D.H.; Chung, H. Experimental study on Venturi-type natural ventilator. *Energy Build.* **2017**, *139*, 232–241. [CrossRef]
11. Kildes, J.; Vallarino, J.; John, D. Spengler, Dust build-up on surfaces in the indoor environment. *Atmos. Environ.* **1999**, 33, 699–707. [CrossRef]
12. ASHRAE. *ANSI/ASHRAE Standard 62.1-2013 Ventilation for Acceptable Indoor Air Quality*; American National Standard: Atlanta, GA, USA, 2013.
13. Kim, T.; Lee, D.H.; Ahn, K. Characteristics of rain penetration through a gravity ventilator used for natural ventilation. *Ann. Occup. Hyg.* **2008**, *52*, 35–44. [CrossRef]
14. Song, S.K. Hiroshi Matsumoto, A study on the natural ventilation performance of a turbine ventilator for houses. *Environ. Eng.* **2010**, *75*, 157–163. [CrossRef]
15. Favarolo, P.A.; Manz, H. Temperature-driven single-sided ventilation through a large rectangular opening. *Build. Environ.* **2005**, *40*, 689–699. [CrossRef]
16. Ghiaus, C.; Allard, F. *Natural Ventilation in the Urban Environment-Assessment and Design*; Earthscan: London, UK, 2005.
17. Hunt, G.R.; Linden, P.P. The fluid mechanics of natural ventilation-displacement ventilation by buoyancy-driven flows assisted by wind. *Build. Environ.* **1999**, *34*, 707–720. [CrossRef]
18. Daisey, J.M.; Angell, W.J.; Apte, M.G. Indoor air quality, ventilation and health symptoms in schools: An analysis of existing information. *Indoor Air.* **2003**, *13*, 53–64. [CrossRef] [PubMed]
19. Santamouris, M.; Synnefa, A.; Asssimakopoulos, M. Experimental investigation of the air flow and indoor carbon dioxide concentration in classrooms with intermittent natural ventilation. *Energy Build.* **2008**, *40*, 1833–1843. [CrossRef]
20. Gan, G.H. Numerical simulation of the indoor environment. *Build. Environ.* **1994**, *29*, 449–459. [CrossRef]
21. Karam, M.; Obaidi, A.; Ismail, M.; Rahman, A. A review of the potential of attic ventilation by passive and active turbine ventilators in tropical Malaysia. *Sustain. Cities Soc.* **2014**, *10*, 232–240.
22. Huang, L.; Ma, J.Y.; Li, A.G.; Wu, Y.Q. Scale modeling experiments of fire-induced smoke and extraction via mechanical ventilation in an underground main powerhouse. *Sustain. Cities Soc.* **2018**, *44*, 536–549. [CrossRef]
23. Gan, G.H. Effective depth of fresh air distribution in rooms with single-sided natural ventilation. *Energy Build.* **2000**, *31*, 65–73. [CrossRef]
24. Gan, G.H. Numerical assessment of thermal comfort and air quality in an office with displacement ventilation. *Build. Environ.* **2003**, *38*, 201–223.
25. Visagavel, K.; Srinivasan, P.S.S. Analysis of single side ventilated and cross ventilated rooms by varying the width of the window opening using CFD. *Sol. Energy* **2009**, *83*, 2–5. [CrossRef]
26. Wright, T.; Simmons, W.E. Blade Sweep for Low Speed Axial Fans. *ASME Turbomach.* **1990**, *112*, 151–158. [CrossRef]
27. Etheridge, D.W. Natural Ventilation through Large Openings—Measurements at Model Scale and Envelope Flow Theory. *Int. J. Vent.* **2004**, *2*, 325–342. [CrossRef]
28. Beiler, M.G.; Carolus, T.H. Computation and measurement of the flow in axial flow fans with skewed blades. *J. Turbomach.* **1999**, *121*, 59–66. [CrossRef]
29. Farrall, M.; Simmons, K.; Hibberd, S. A numerical model for oil film flow in an aero-engine bearing chamber and comparison with experimental data. In Proceedings of the ASME Turbo Expo 2004: Power for Land, Sea, and Air, Vienna, Austria, 14–17 June 2004; pp. 110–119.
30. Elgowainy, A.; Feinstein, D.; Rodriguez, I. Performance analysis of axial fans in unitary air-conditioning systems. In Proceedings of the IECEC '02. 2002 37th Intersociety Energy Conversion Engineering Conference, Washington, DC, USA, 29–31 July 2002; pp. 531–536.
31. Yang, L.I.; Liu, J.; Ouyang, H. Internal flow mechanism and experimental research of low pressure axial fan with forward-skewed blades. *J. Hydrodyn.* **2008**, *20*, 299–305.
32. Vad, J. Forward blade sweep applied to low-speed axial fan rotors of controlled vortex design: An overview. *J. Eng. Gas Turbines Power* **2013**, *135*, 601–609. [CrossRef]
33. Bleier, F.P. *Fan Handbook Selection, Application and Design*; McGraw-Hill: New York, NY, USA, 1998.

34. Rossetti, A.; Ardizzon, G.; Pavesi, G.; Cavazzini, G. An Optimum Design Procedure for an Aerodynamic Radial Diffuser with Incompressible Flow at Different Reynolds Numbers. *Proc. Inst. Mech. Eng. Part A J. Power Energy* **2010**, *224*, 69–84. [CrossRef]
35. Orosa, J.; Montgomery, M. Compressor Blade with forward Sweep and Dihedral. US Patent US20100054946 A1, 4 March 2010.
36. Sasaki, T.; Breugelmans, F. Comparison of Sweep and Dihedral Effects on Compressor Cascade Performance. *J. Turbomach.* **1998**, *120*, 454–463. [CrossRef]
37. Mingotti, N.; Chenvidyakarn, T.; Woods, A.W. The fluid mechanics of the natural ventilation of a narrow-cavity double-skin facade. *Build. Environ.* **2011**, *46*, 807–823. [CrossRef]

fire

Article

A Study on the Evacuation Spacing of Undersea Tunnels in Different Ventilation Velocity Conditions

Wei Na * and Chen Chen

School of Environment and Energy Engineering, Beijing University of Civil Engineering and Architecture, Beijing 100044, China; chenchen@stu.bucea.edu.cn
* Correspondence: nawei@bucea.edu.cn

Abstract: Ventilation velocity conditions may affect the smoke diffusion and evacuation environment in a tunnel fire, which should be fully considered in evacuation spacing designs of undersea tunnels. This study focuses on reasonable evacuation spacing under various possible velocity conditions of an undersea tunnel, providing a design method reference for calculating safe evacuation spacing. Fire Dynamic Simulator and Pathfinder software were used for numerical simulations of a 50 MW fire and evacuation process in a full-scale undersea tunnel with traffic congestion. The simulation cases contained velocity modes from zero to satisfying the critical velocity and evacuation spacings from 30 m to 80 m. The calculated distributions of the available safe escape time indicated that a low ventilation velocity, such as 1.0 m/s, is beneficial to the downstream evacuation, but turning off mechanical ventilation increased risk near the fire source. The required safe escape time is shortened with a reduction in slide spacings, but the shortened rate slowed down after spacing was less than 60 m. In addition, the slow evacuating areas from 100 m to 300 m from the fire source independent of spacing are identified. Ultimately, the reasonable evacuation spacings of 60 m, 50 m, and 30 m, corresponding to three possible ventilation velocity modes of low, medium, and high, are proposed through the comparisons of the distributions of available safe escape time and required safe escape time.

Keywords: undersea tunnel fire; evacuation spacing; longitudinal ventilation velocity; numerical simulation; ASET; RSET

Citation: Na, W.; Chen, C. A Study on the Evacuation Spacing of Undersea Tunnels in Different Ventilation Velocity Conditions. *Fire* **2022**, 5, 48. https://doi.org/10.3390/fire5020048

Academic Editors: Chuangang Fan and Dahai Qi

Received: 10 March 2022
Accepted: 4 April 2022
Published: 7 April 2022

1. Introduction

Underwater tunnels, especially undersea tunnels, are usually considered as essential components in urban traffic engineering due to their heavy traffic volumes, complex vehicle composition, obvious slope, and consequent high fire risk with complicated heat transfer [1] and, hence, mostly have larger scales as well; thus, the shield method is widely used in their construction. Moreover, fewer installations of transverse passages and escape shafts in undersea tunnels [2] are harmful to fire evacuation safety because of the limitations of the structure and internal space. Therefore, reasonable design alternatives for ventilation and evacuation are promising steps to reduce initial and operational costs and to improve the long-term operational performance for undersea tunnels [3].

The high construction cost of evacuation passages and smoke vents for an undersea tunnel is attributed to the complex geological environment; thus, full jet longitudinal ventilation and longitudinal evacuation with bottom passage are adopted in a large proportion of undersea tunnels. Unlike the transverse evacuation method using cross passages or using a parallel escape tunnel between two tunnels, longitudinal evacuation is possible with simple construction and low cost. One only needs to install evacuation slides or stairs at a reasonable spacing inside the tunnel to connect the upper space and the bottom passage, avoiding adding auxiliary facilities outside. Related simulation work [4] found that the evacuation spacing of longitudinal evacuation was 150~200 m shorter than that of transverse evacuation on the conditions of equivalent evacuation capacity and satisfying evacuation safety. The denser evacuation spacing means that more evacuation slides (stairs) need to be built, which requires a balance between engineering economy and evacuation safety. A basic principle for ensuring life safety in tunnel fires is that the available safe escape time (ASET) is greater than the required safe escape time (RSET) by an adequate safety margin [5]. Too large a spacing may mean that the RSET does not satisfy evacuation safety, while too small a spacing leads to an excessive safety margin and a high engineering cost. Thus, reasonable evacuation spacing designs are crucial to balance economy and safety.

A lack of consistent standards is a threat to reasonable fire evacuation design for undersea tunnels. Fire smoke is the biggest threat to the people trapped [6], and longitudinal ventilation velocity can significantly affect the hazard level of the smoke environment and then affect the ASET. The high-velocity mode is recommended by the Chinese design standard for ventilation of highway tunnels [7] for the operation of fire ventilation in tunnels with one-way traffic; i.e., the longitudinal ventilation velocity is equal to or greater than the critical velocity (v_c) to eliminate the smoke back-layering occurring in the tunnel. The related design standards of Germany, France, Switzerland, Austria [8,9] and the Permanent International Association of Road Congress (PIARC) [10] proposed that the low-velocity mode could be adopted in the self-evacuation phase after fire, which requires the longitudinal velocity to be far lower than v_c in order to maintain the stability of smoke stratification downstream of the fire source. In the meanwhile, various perspectives have been presented by other authors, including Xue et al. [11], who recommended that the ventilation velocity should be lower than or close to v_c. Hu [12] and Guo et al. [13] suggested using stratified velocity and constrained velocity, respectively. Liu [14] suggested that the velocity should be regulated below 1.0 m/s. Chen et al. [2] posited that mechanical ventilation should be turned off during evacuation. Jan et al. [15] and Espinosa et al. [16] believed that using two-stage ventilation (low velocity to high velocity) is more conducive to smoke control and evacuation in undersea tunnel fires. In other words, during the ventilation control stage of a tunnel fire, the decision makers or control system may use any velocity modes based on the judgment of fire scenes to achieve different requirements, such as ensuring no smoke in upstream rescue routes, or maintaining the downstream smoke stratification as much as possible, or using a combination of these methods. Fully considering the possible ventilation conditions is very meaningful for evacuation spacing design. Yin et al. [17] considered the ineffective and effective situations of exhaust, then proposed a safety spacing of 50 m and 75 m accordingly, by calculating the ASET distributions of two conditions. In most previous studies of evacuation spacing design, however, empirical fixed values of ASET were usually used to simplify the calculation, ignoring the influence of the possible conditions of ventilation velocity on the distribution of ASET. Yuan et al. [18] experimented and calculated the evacuation process of an undersea tunnel with longitudinal bottom evacuation, referring to the empirical value 360 s of ASET, and proposed that the reasonable value of evacuation interval was 64.3 m. Zhang et al. [19] and Shen et al. [20] put forward a reasonable slide spacing of 80 m for different undersea tunnels in Chittagong and Hangzhou by using ASET values of 900 s and 600 s, respectively, according to engineering experiences. Hui et al. [21] used an ASET value of 480 s and recommended a slide spacing of 45 m. The conclusions of the above studies were not comprehensive, which indicates the limitations of using empirical fixed values. Reasonable evacuation spacing designs need to rely more on accurate analyses of ASET distributions. Thus, it is necessary to take into account the dynamic distribution of ASET in different ventilation velocity conditions of tunnel fire and make corresponding design suggestions.

The present study concentrated on the reasonable evacuation spacings of undersea tunnels in possible longitudinal ventilation conditions. The ventilation and smoke diffusion of tunnel fire were dynamically simulated by FDS(V6). The effects of five typical ventilation velocities of four velocity modes ranging from 0 m/s to satisfying v_c on ASET distributions were investigated, and the RSET distributions were calculated through the simulations of evacuation processes in six spacings by Pathfinder software. Ultimately, design suggestions of reasonable spacings in different ventilation velocities that satisfy evacuation safety requirements were put forward through the comparative analysis of ASET with RSET, thereby providing a design method reference for calculations of evacuation spacing in undersea tunnels.

2. Numerical Modeling

2.1. Tunnel Descriptions

2.1.1. Tunnel Configuration

A medium-length, double-hole undersea tunnel, built using tunneling shields in the city of Macao, was selected in this study. The tunnel is a one-way, two-lane tunnel with a shield part 1000 m long, 12 m wide, and 8 m high, in the shape of a horseshoe for the space above the road with a sectional area of 71.7 m^2, without cross passages between the two tunnel holes. The road near the inlet and the outlet is sloped. The schematic diagram of the shield part of the tunnel is shown in Figure 1, where Y = 0 m denotes the entrance of the shield part, while Y = 1000 m denotes the exit of the same.

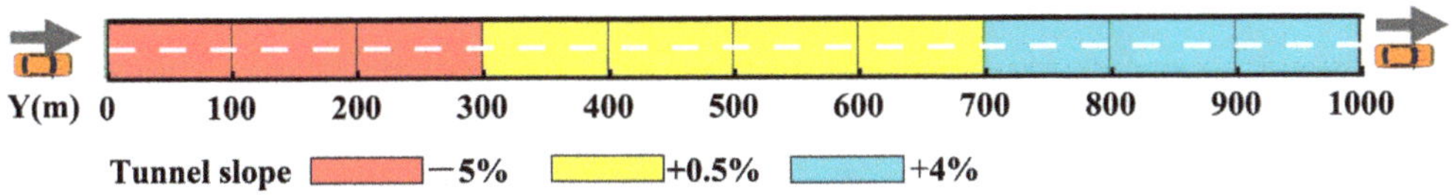

Figure 1. Schematic diagram of the shield part of the tunnel.

2.1.2. Ventilation Conditions

The tunnel adopts a full-jet longitudinal ventilation system. To focus on the influence of ventilation velocity conditions on ASET, only the velocity value was considered, and other factors affecting smoke diffusion were ignored. Due to the large range of possible ventilation velocities in actual situations of tunnel fire, extensive selections of velocity conditions would make the total workload too large. Therefore, the ventilation velocity was simply divided into several modes, and the typical velocity values of each mode were set according to the ventilation velocity studies mentioned above. The velocity modes include "turn off mechanical ventilation," "low-velocity," "medium-velocity," "high-velocity (close to the critical velocity v_c)," and "high-velocity (satisfying the critical velocity v_c)". Typical velocity values are shown in Table 1, which are the average air velocities in the tunnel after the stable operations of jet fans.

Table 1. Different velocity modes and values in tunnel fire.

Velocity Mode	Value (m/s)
Turn off mechanical ventilation	0.0
Low-velocity	1.0
Medium-velocity	2.0
High-velocity (close to v_c)	4.5
High-velocity (satisfying v_c)	6.0

2.1.3. Evacuation Conditions

The tunnel adopts a longitudinal evacuation method with a bottom passage linked by evacuation slides. Evacuation slides are installed from the middle position (Y = 500 m) to both sides of the tunnel equidistantly. According to the engineering experience range of slide spacing in tunnels, six schemes of different slide spacings from 30 m to 80 m were proposed, as shown in Table 2. The priority was to satisfy the evacuation spacings in the middle of the tunnel. Due to the high evacuation capability at the entrance and exit of the tunnel, the spacing between the slides on the edge and the entrance (exit) is slightly larger.

Table 2. Schemes of tunnel slide spacing.

Slides Spacing (D)	Number of Slides
30 m	31
40 m	23
50 m	19
60 m	15
70 m	13
80 m	11

2.2. Modeling and Validation

2.2.1. Physical and Mathematical Model

A full-scale 3D model of the tunnel was constructed by PyroSim software and simplified vehicle entities in the whole tunnel were built to correspond to reality situations, as shown in Figure 2. The calculation kernel of PyroSim is based on FDS (version 6.7.5), developed by the National Institute of Standards and Technology (NIST). FDS numerically solves a form of the Navier–Stokes equations appropriate for low speeds ($Ma < 0.3$), thermally driven flow with an emphasis on smoke and heat transport from fires [22].

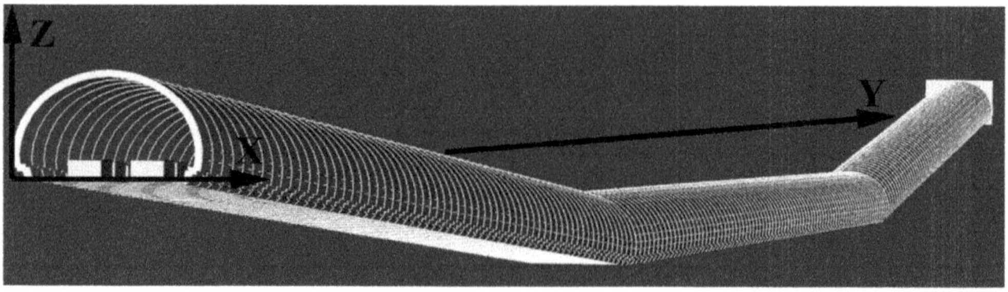

Figure 2. Schematic diagram of the full-scale 3D physical model of tunnel.

In order to highlight the influences of ventilation velocity on smoke diffusion and ASET distributions, parameters such as installation position and the start sequence of the fans are ignored. The airflow of each ventilation condition was provided by the air-supply surface of constant flow outside the tunnel entrance, and the feasibility refers to the experimental studies of Tang et al. [23] and Chow et al. [24]. ASET is defined as the time interval between the time of ignition and the time passengers are estimated to be incapacitated [25], and it is also the tenability limit calculated depending on the hazard evaluation of the smoke environment. Carbon monoxide (FED_{CO}), cumulative heat (FED_{heat}), the temperature of convective heat (Temp), and evacuation visibility (Vis) were used as indexes to calculate ASET in this study, as shown in Table 3.

We used a visibility threshold of 3 m because of the dense slides in the bottom evacuation mode. Most of the passengers could quickly reach the vicinity of the evacuation slide before the visibility drops significantly, and then they can keep evacuating until they would turn back than rather enter the smoke.

The sensors and slices were set up to obtain the CO concentration, temperature, air velocity, and visibility data of the evacuation space for monitoring ambient air velocity and recording data for the above indexes. The 2 m height of installed sensors was based on the range of evacuating activity, considering taller individuals, and leaving a safety margin to ensure that smoke hazards would not be underestimated. The sensors and slices for simulation are shown in Table 4.

Table 3. Calculation index of available safe escape time (ASET).

Index	Symbol	Thresholds	Detail
The maximum fractional effective dose of carbon monoxide	FED_{CO}	0.3	Accumulated CO inhalation exceeds the threshold value resulting in passengers being incapacitated [26], considering the decline of tolerance threshold of people in exercise [27]
The maximum fractional effective dose of heat	FED_{heat}	0.3	Accumulated heat exceeds the threshold value resulting in passengers being incapacitated, which includes convective heat part and radiant heat part [26]
Threshold temperature of convective heat	Temp (°C)	60	Heat burns the respiratory tract [26]
Visibility	Vis (m)	3	30% of people turn back rather than enter the smoke [5,28].

Table 4. Sensors and slice settings.

Type	Detail	Number	Location
Gas-phase sensors	convection temperature radiant heat flux CO volume fraction	47 47 47	Longitudinally arranged at 2 m above the road every 10~40 m along X = 6 m
	air velocity	21	Longitudinally arranged at 4 m above the road every 20~50 m along X = 6 m
Slice of parameter distribution	temperature field velocity field CO concentration field visibility field	2 2 2 2	Arranged at the position of X = 6 m and X = 10 m, covering tunnel length.

The initial parameters of model boundary conditions were set as follows:

(1) In the initial stage, the pressure inside the tunnel was P = 101 kPa, each velocity component was zero, and the air temperature was T = 20 °C;
(2) Walls and pavements were concrete adiabatic surfaces without internal heat sources and a friction coefficient $\lambda = 0.02$;
(3) The surface of the retained vehicle in the tunnel was adiabatic;
(4) Tunnel exits and entrances were open surfaces, with pressure exits connected to the atmosphere.

The computational domain of the model was made up of 15 adjacent linked meshes and covers the entire tunnel. To shorten the simulation time, multi-core parallel computing of a single CPU was carried out. The grid and simulation settings are shown in Table 5.

Table 5. Grid and simulation settings.

Parameter	Details
Computational domain (m)	$0 \leq X \leq 12$, $-5 \leq Y \leq 1005$, the Z direction computational domain is set flexibly according to the road slope.
Grid size (m)	uniformly selected 0.4 × 0.4 × 0.4 cube grid in each mesh.
Total number of grids	2,493,760
Processor parameters	16 cores (4.0 GHz)
OpenMP Threads MPI	15
Simulation type	large eddy simulation (LES)
Eddy viscosity	0.1
Near-wall eddy viscosity	0.6
Time step	system default value without limiting time step

Pathfinder software, which has been commonly used in the field of fire evacuation [29,30], was used to construct the evacuation model in this study and was used in steering mode to consider the passenger collisions [31]. According to the traffic flow data that will be mentioned in Section 2.2.2, assuming that all blocked vehicles are randomly distributed and the distance between vehicles is 1 m, a total of 310 vehicles and 1870 passengers could possibly be trapped in the tunnel. The average shoulder width of passengers was set to 0.5 m, in line with the relevant fire safety standard. Evacuation slides on the side walls of the tunnel were simplified as doors to the outside. In addition, passengers could flexibly choose to evacuate from the evacuation slide or tunnel entrance according to the distance or flow without crossing the fire source. The partial evacuation model is shown in Figure 3. The white closed lines denote vehicle outlines, the green points denote passengers, and the short green line below denotes an evacuation slide.

Figure 3. Partial view of evacuation model.

2.2.2. Operation Modes and Fire Scenarios

Traffic Modes

As an extremely unfavorable situation that may occur, the condition of traffic congestion in the entire tunnel was assumed in this study. In this case, the number of vehicles and the number of people to be evacuated are the largest. The traffic composition of the tunnel, as shown in Table 6, was based on the vehicle-type distribution data of the roads near the tunnel, which were obtained on the Macau DSAT traffic video monitoring platform (http://www.dsat.gov.mo, accessed on 15 August 2021), and the average passenger number of each vehicle was assumed from experience data.

Table 6. Traffic composition of tunnel.

Vehicle Types	Proportion (%)	Average Passenger Number
Sedan car	55	4
MPV	20	6
Small Bus	3	20
Bus	8	45
HGV	14	2

Evacuation Modes

The evacuation process can be divided into two parts, the ready process and the travel process, which could be reflected in the composition of RSET. RSET is defined as the calculated time period required for an individual passenger to travel from their location at the time of ignition to a safe refuge or place of safety [25], which consists of the detect time, alarm time, pre-travel activity time, and travel time, as shown in Figure 4. For simplicity, the ready time (RT) is considered to be the time period required for the ready process.

In the design stage, it is difficult to determine a clear safety margin, so the basic requirement ASET $\geq$ RSET was used as the safe evacuation condition, which was also adopted in the research of Song et al. [32], and Zhang et al. [33]. During the ready process, passengers close to the fire will start traveling immediately after the ignition, and RT is zero. However, passengers far from the fire need to wait for the arriving crowd or the ringing alarm to begin to move; similar to a domino effect, their RT goes up because the moving crowd or diffusing smoke will prompt others to realize the risk. Considering the prompt effect of smoke diffusion and crowd movement on evacuation, the RT was set according to the distance from the fire source, as shown in Table 7.

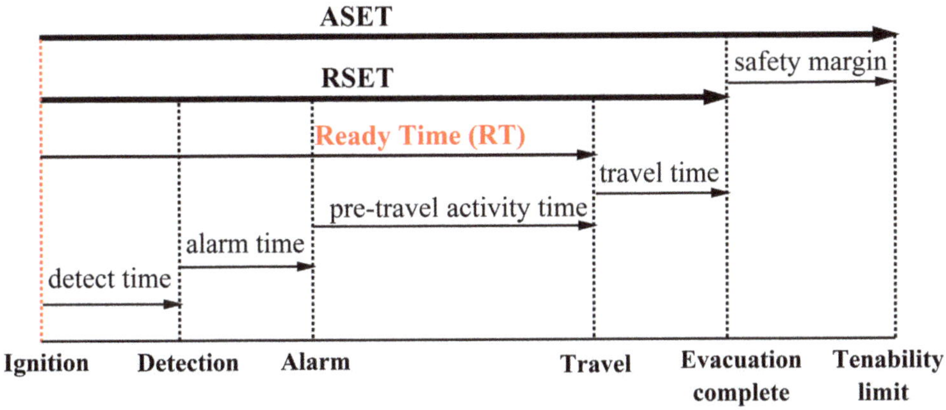

Figure 4. The relationship between the available safe escape time (ASET) and the required safe escape time (RSET) in evacuation process.

Table 7. RT of evacuation in different distance from the fire source.

Distance from the Fire Source S(m)	RT (s)
$S \leq 20$	0
$20 < S \leq 60$	30
$60 < S \leq 120$	60
$120 < S \leq 200$	90
$S > 200$	120

The maximum walking speed of passengers may undergo complex changes influenced by the smoke environment and psychological fluctuation during the travel process, which is not the focus of this article. To simplify this, referring to the study of Shen et al. [20], Seike et al. [34], and the average walk speed given by PIARC [35], the maximum values of walking speed were set as 1.0 m/s in the horizontal part and 0.8 m/s in the slope part, which represent a low level in adverse situations and make sure that the RSET calculations would not be underestimated. The evacuation capacity of each slide was set as 24 person/min, referencing the experimental and simulation research data from Xie et al. [36], Zhang et al. [37], and Cao [38]. There is no upper limit for the evacuation capacity of the tunnel's entrance and exit.

Fire Scenarios

It is reasonable to assume that the cause of the tunnel fire is a heavy goods vehicle (HGV) accident, considering that this type of accident is responsible for the highest proportion of vehicle fire accidents in China [39]. Additionally, the accident in the Mont Blanc tunnel can also prove the serious consequences of HGV fires [40]. The peak heat release rate (HRR) of the fire was set as 50 MW in accordance with [41]. The fire growth category was defined as ultra-fast (coefficient $\alpha = 0.18$), increasing by the function t^2; HRR peaks at 530 s after the fire. The accident HGV was located in the center of the tunnel at X = 6 m, Y = 500 m. To lower the impact of fire scale changes on ASET, fire spread between vehicles was not considered. The model of the fire source used a burner surface with a size of 1 m × 2 m (X × Y) and 1 m above the road, which was consistent with the HGV carriage location. The combustion of the fire was specified as polystyrene (PS), which is a widely used insulation material. The byproduct parameters including the soot yield and the CO yield were specified as 0.16 and 0.06, respectively, drawing on the empirical data [42].

2.2.3. Grid Independence Verification and Model Validation

Grid Independence Verification

Grid independence verification was performed to evaluate the dependency of fire simulation results pertaining to the size and number of grids. Appropriate grid size can be accessed by the dimensionless expression D^*/δ_x [22], where δ_x is the grid size (m), and D^* is the characteristic diameter of the fire source (m), as shown in Equation (1). For accurate simulation results to be obtained, it is recommended that D^*/δ_x be between 4 and 16.

$$D^* = \left(\frac{\dot{Q}}{\rho_\infty c_p T_\infty \sqrt{g}}\right)^{\frac{2}{5}} \quad (1)$$

Here, $\dot{Q}$ is the peak heat release rate of fire (kW); ρ_∞ is the ambient air density (kg/m^3); c_p is the ambient air specific heat (kJ/(kg·K)); T_∞ is the ambient air temperature (K); and g is the gravity acceleration (m/s^2).

D^* of the 50 MW fire was calculated to be 4.6 m, so the grid size should be between 0.3 m and 1.1 m. The results of independent verification for different grid sizes are compared in Table 8. The absolute percentage errors (APE) of temperature values on both sides of the fire source were calculated, with the temperature detection points set at positions of 7.5 m height from the road and along the central axis in the Y direction. The APE was less than 1.8% between the 0.4 m grid and the 0.3 m grid. There was no significant difference in temperature distributions if the grid size was less than 0.4 m, so it was acceptable to use a 0.4 m grid size for the model of the tunnel fire in Section 3.1. In addition, it could be found that a more accurate description of the fire plume can be obtained by using a smaller grid size near the fire source, such as 0.2 m, and a significantly increased computation time. Considering that its effect on the smoke parameters in the entire tunnel is very small, for the purpose of this work, did not use dense grids near the fire source.

Table 8. Independent verification of grid size.

Grid Size (m)	Absolute Percentage Error (APE) (%)			
	−25 m	−10 m	+10 m	+25 m
0.3	\	\	\	\
0.4	1.8	1.8	0.3	0.8
0.5	6.6	4.1	2.7	4.0

Model Validation

Since it is difficult to find experimental works with a similar fire scale and similar tunnel structure to this work, to verify the reliabilities of the simulation software and the grid calculation method in this study, using the same method of modeling, boundary condition setting, and grid calculating, to simulate a 0.75 MW combustion experiment in an 88 m-long channel carried out by Hu et al. [43]. The full-scale FDS model included the same tunnel structure, fire location, HRR, and the same sensor placements. The purpose of this section is to prove the feasibility of the software and modeling methods used, rather than to completely replicate the experiment, so the byproduct parameters related to the fuel type are ignored, and only the ceiling temperature data from the simulation and the experiment were compared. The ceiling temperature can comprehensively reflect the smoke diffusion and heat transfer process, and it can be regarded as a characteristic parameter for evaluating similar combustions. The ceiling temperature distributions within 80 m of the fire source were compared, as shown in Figure 5. The simulated data by FDS are represented by the red line, and the experimental data from Hu et al. are shown as black cubes.

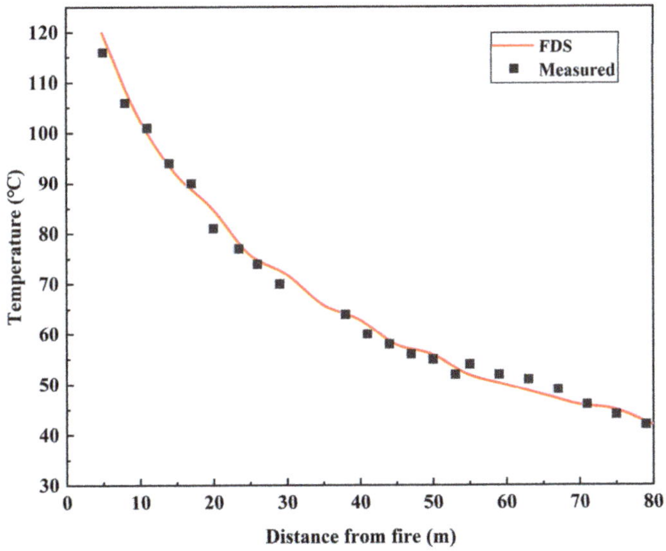

Figure 5. Comparison of temperature distributions between simulation and experiment.

It could be found that the distribution data simulated by FDS were highly coincident with the experimental measurement data, with the 2.5% mean absolute percentage error (MAPE), proving that the grid calculation methods and the FDS simulation used above are reliable. The reliability of the model mentioned in Section 3.1 was demonstrated.

3. Result and Discussion

3.1. Longitudinal Ventilation Velocity vs. Available Safe Escape Time

The influence of longitudinal ventilation velocity on ASET was discussed in this section. Distributions of ASET and back-layering length (BLL) at different longitudinal ventilation velocities of 0 m/s, 1.0 m/s, 2.0 m/s, 4.5 m/s, and 6.0 m/s are shown in Figure 6. The following results were obtained:

(1) The risk of the smoke environment in the downstream tunnel is higher than that in the upstream tunnel. In the case of turning on mechanical ventilation, the average ASET values downstream (Y > 500 m) of the fire source were significantly lower than those upstream (Y < 500 m), with a reduction of 21% to 65% at the identical ventilation velocity. In the case of turning off the mechanical ventilation, it could be seen that the distribution of upstream and downstream ASET was not symmetrical. The reason for this is that the slopes of this tunnel are different on both sides of the tunnel, and the hot pressure releases preferentially by the exit of the downstream tunnel due to different chimney effects. Thus, the diffusion range of upstream smoke was small and the upstream ASET distribution was higher.

(2) Turning off the mechanical ventilation increases the environmental risk on both sides of the fire source compared with the low-velocity mode of 1.0 m/s. If the ventilation velocity is 0 m/s, the ASET values in the area of 250 m upstream and downstream of the fire source will be substantially lower than that at the velocity of 1.0 m/s. This might be because the high concentration of smoke accumulated near the fire source, resulting in a rapid decline of visibility, due to the lack of airflow control, which hampers the upstream rescue, so it is not recommended. In addition, without the cooling effect of ventilation, airflow may also cause the excessive temperature at the ceiling, which may damage the tunnel structure.

(3) Higher ventilation velocities increased the evacuation risk compared with the low-velocity mode of 1.0 m/s. Although the BBL of 1.0 m/s velocity was the longest, it

has almost no threat to upstream passengers, which is reflected in the ASET, because the back-layering with low concentrations did not sink into the evacuation space. The higher the ventilation velocity, the lower the distributions of ASET. For the low-, medium-, and high-velocity modes, the average ASET value decreased by 8% to 25% for each 1 m/s increase in ventilation velocity. The ASET value decreased slightly and tended to be constant in the process of velocity increases from 4.5 m/s to 6.0 m/s, which was almost the lowest distribution of ASET. It can be inferred from the above that the lower ASET value at a higher ventilation velocity might be related to the destruction of smoke stratification. To prove this hypothesis, the time of smoke sinking into the evacuation space of 2 m height from the road within 150 m downstream of the fire source at longitudinal ventilation velocities of 0 m/s, 1.0 m/s, and 2.0 m/s is illustrated in Figure 7. The stratification was not affected by ventilation airflow at a velocity of 0 m/s. Although accurately figuring out the duration of stable smoke stratification was difficult, it was straightforward to see the relationship between stratification and velocity. The average times for smoke flow sinking into the evacuation space of 1.0 m/s and 2.0 m/s velocities were 67% and 41% of the average time at the velocity of 0 m/s. With the increase in velocity, the smoke stratification downstream was more unstable due to more airflow disturbances and would sink into the evacuation space earlier. The sinking smoke would bring heat, CO, and low visibility to passengers, and result in low ASET values.

(4) The computation of ASET mainly depends on high convective heat and low visibility. In the process of selecting the index that first reached the threshold value to compute the ASET, it was found that the temperature index close to the fire source first reaches its threshold value, and the rapid reduction in visibility is the primary threat faced by passengers in other places, which is also consistent with the conclusion of Gehandler et al. [44]. Take a ventilation velocity of 2 m/s as an example, as shown in Figure 8. In addition, the heat and CO from hot toxic smoke are not the key factors for determining ASET, because they reach the threshold very slowly due to the action of buoyancy and ventilation airflow.

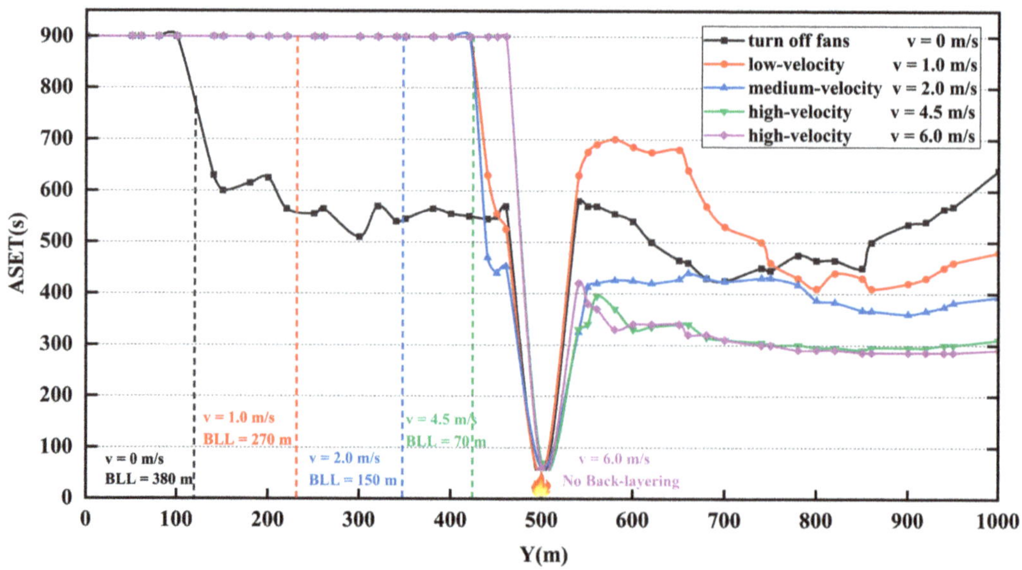

Figure 6. Distribution of ASET and back-layering length (BLL) at different longitudinal ventilation velocity (v).

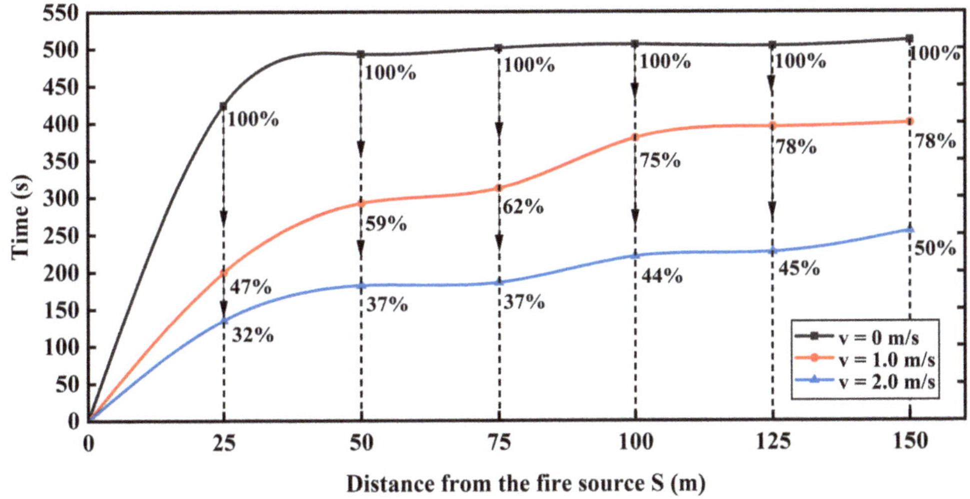

Figure 7. Time of downstream smoke sinking into evacuation space (within 2 m height from the road) after ignition.

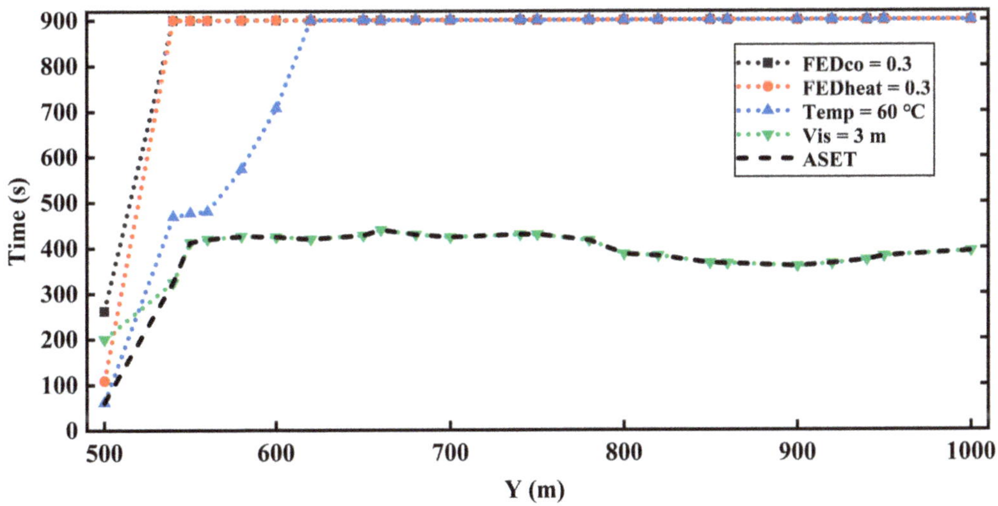

Figure 8. The downstream ASET distribution and the time when FED_{CO}, FED_{heat}, temperature, and visibility indexes reach the threshold at the velocity of 2.0 m/s.

3.2. Evacuation Slides Spacing vs. Required Safe Escape Time

The influence of spacing and number of evacuation slides on RSET in different longitudinal ventilation velocities are discussed in this section. The RSET distributions on both sides of the fire source were similar because slides were symmetrically installed. The RSET distributions under conditions of different slide spacings (D) are shown in Figure 9. The RSET distributions of the downstream tunnel (Y > 500 m) when using different D values from 30 m to 80 m are shown as Curve 1, and the red dots denote the RSET values near the evacuation slides. The variation in the longest evacuation times for the entire tunnel (the max RSET values in Curve 1 for increasing D) are recorded as Curve 2.

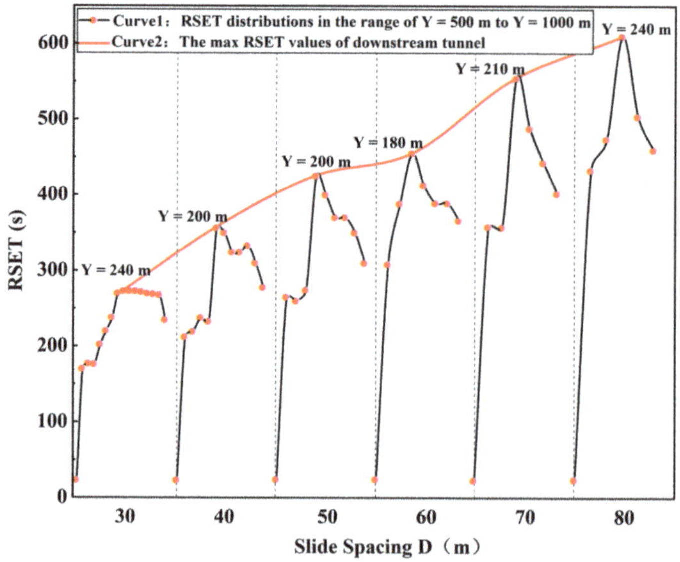

Figure 9. RSET distributions in conditions of different slide spacing (D).

It could be found that with the decrease in slide spacing D, the downstream RSET showed a nonlinear downward trend, and the inflection point might be at D = 60 m. To be specific, as D reduced from 80 m (total 11 slides) to 60 m (total 15 slides), the max RSET value shortened by 39 s for each additional slide. As D continued reducing from 60 m to 30 m (total 31 slides), the max RSET value shortened 11 s for each additional slide, and the shortened time was only 28% of the former. This indicated that the effect of decreasing RSET by further reducing the evacuation slide spacing (increasing the number of slides) is weakened after D is less than 60 m. It could also reflect the fact that the cost-effectiveness of installing slides declines after the slide spacing is less than 60 m.

Further, to identify the slow evacuating area and the high-risk area of the tunnel, the tunnel was longitudinally divided into five areas, I, II, III, IV, and V, by the distance from the fire source. Each area had two parts, including a 100 m upstream part and a 100 m downstream part, as shown in Table 9. Taking the low-velocity ventilation mode as an example, the comparison of RSET distributions with the ASET distribution at the velocity of 1.0 m/s is shown in Figure 10. The ASET curve is the actual ASET distribution of the tunnel; the average ASET curve is the average ASET value in each part, which roughly reflects the overall situation in the part. The RSET curve represented the variation of the maximum RSET values when D increased from 30 m to 80 m in each part, the leftmost point represented the condition of D = 30 m, and each RSET value belonged to the passenger with the longest evacuation time in each part.

Table 9. Area-dividing method.

Area	Distance from the Fire Source S (m)	Upstream Part Y (m)	Downstream Part Y (m)
I	$S \leq 100$	$400 \leq Y < 500$	$500 < Y \leq 600$
II	$100 < S \leq 200$	$300 \leq Y < 400$	$600 < Y \leq 700$
III	$200 < S \leq 300$	$200 \leq Y < 300$	$700 < Y \leq 800$
IV	$300 < S \leq 400$	$100 \leq Y < 200$	$800 < Y \leq 900$
V	$400 < S \leq 500$	$0 \leq Y < 100$	$900 < Y \leq 1000$

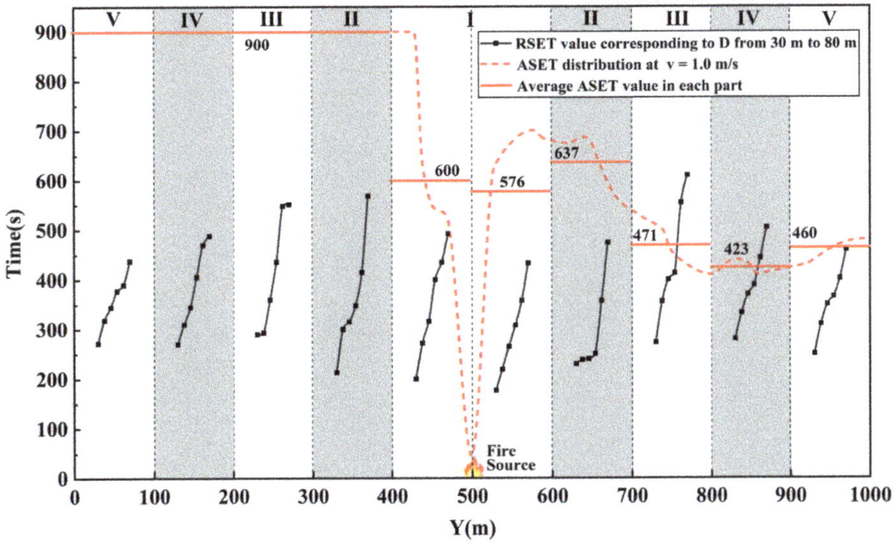

Figure 10. ASET compared with RSET corresponding to different D under the condition of v = 1.0 m/s.

As can be seen in Figure 10, the slow evacuating areas were II and III, independent of the value of D. The slow evacuating areas had a larger overall RSET value than other areas, which was always from 100 m to 300 m from the fire source. It could also be proved by Figure 9 that with the increasing distance from the fire source, the RSET value firstly increased and then decreased, and the extreme point was always in these areas. The main reason for this was that passengers in area I had a shorter RT and the crowd moved earlier to both sides of the tunnel after ignition. This increased the evacuation pressure on areas II and III, but this effect gradually reduced in areas IV and V far from the fire source.

The high-risk evacuating areas were the downstream parts of III and IV when $D \geq 70$ m. Figure 10 shows that in the downstream parts of 200 m to 400 m from the fire source, the average ASET values were lower than the RSET values under conditions of D = 70 m and D = 80 m. However, with the change in ventilation velocity and ASET distributions, the high-risk areas expand. It could be stated at least that areas III and IV were always in danger when $D \geq 70$ m, regardless of the ventilation velocity because the ASET distributions of other velocities were all lower than that of 1 m/s, as shown in Figure 6.

To explore the reasonable D in conditions of different ventilation velocities that could possibly occur, the downstream ASET distributions of the low-velocity mode, medium-velocity mode, and high-velocity mode were compared with the downstream RSET distributions of different evacuation spacing Ds from 30 m to 70 m, as shown in Figure 11. The RSET columns recorded the longest evacuation time of each slide, and the evacuation assessments near the evacuation slides represented the worst situation of the entire tunnel. If the ASET distribution of one velocity mode was always higher than the RSET value of one spacing D at any position, it could be considered that the evacuation process was secure, and the maximum D value satisfying the above requirement was the reasonable evacuation spacing of that velocity mode.

Further, to quantify the safety of each evacuation schemes, Table 10 counts the number of trapped people (among the total of 1870 passengers involved) corresponding to different evacuation spacings under possible ventilation modes. The trapped passengers are located in the position where ASET is less than RSET, as shown in Figure 11. The relative value of the number of trapped passengers can also be reflected by the difference value of time. In each velocity mode, as the evacuation spacing shortens, the number of trapped passengers gradually decreases. Under the same evacuation spacing scheme, the number of trapped passengers is ranked as: low-velocity mode < medium-velocity mode < high-velocity mode, which can indicate that using the low-velocity mode is beneficial to safe evacuation. Specifically, the reasonable slide spacing should not be larger than 60 m if using the low-velocity ventilation mode of 1.0 m/s, it should not be larger than 50 m if using the medium-velocity mode of 2.0 m/s, and should not be larger than 30 m if using the high-velocity mode above 4.5 m/s. The reasonable spacings recommended above do not represent the accurate value but only acceptable ranges. This is because the difference between the evacuation spacing schemes selected in this study is 10 m, which conforms to the general engineering design accuracy.

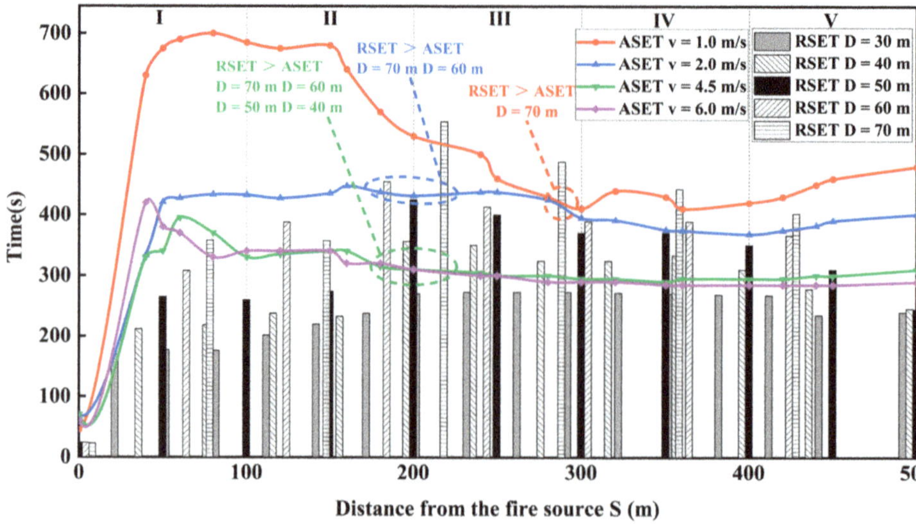

Figure 11. Comparison of RSET of different D with ASET of different velocities in the downstream area of fire source within 500 m.

Table 10. Number and location of trapped passengers under different evacuation spacings.

Evacuation Spacing D (m)	Number and Location of Trapped Passengers in Tunnel (among the 1870)			
	Low-Velocity Mode	Medium-Velocity Mode	High-Velocity Mode	
	1.0 m/s	2.0 m/s	4.5 m/s	6.0 m/s
30	0	0	0	0
40	0	0	79 (19 at 700 m, 18 at 740 m, 10 at 780 m, 12 at 820 m, 14 at 860 m, 6 at 900 m)	94 (19 at 700 m, 20 at 740 m, 14 at 780 m, 14 at 820 m, 17 at 860 m, 10 at 900 m)
50	0	0	174 (46 at 700 m, 41 at 750 m, 30 at 800 m, 32 at 850 m, 22 at 900 m, 3 at 950 m)	181 (46 at 700 m, 41 at 750 m, 31 at 800 m, 34 at 850 m, 26 at 900 m, 3 at 950 m)
60	0	13 (8 at 680 m, 5 at 374 m)	220 (21 at 620 m, 57 at 680 m, 43 at 740 m, 36 at 800 m, 36 at 860 m, 27 at 920 m)	227 (19 at 620 m, 55 at 680 m, 45 at 740 m, 38 at 800 m, 40 at 860 m, 30 at 920 m)
70	42 (14 at 710 m, 23 at 780 m, 5 at 850 m)	111 (47 at 710 m, 25 at 780 m, 27 at 850 m, 12 at 920 m)	284 (8 at 640 m, 98 at 710 m, 73 at 780 m, 61 at 850 m, 44 at 920 m)	297 (3 at 570 m, 8 at 640 m, 99 at 710 m, 77 at 780 m, 63 at 850 m, 47 at 920 m)

4. Conclusions

The reasonable evacuation spacings of undersea tunnels in possible longitudinal ventilation conditions were researched in this work. FDS and Pathfinder were used for fire and evacuation simulations of a 50 MW fire in a full-scaled tunnel with traffic congestion. The effects of five typical ventilation velocities on ASET distributions and the effects of six evacuation spacings on RSET distributions were analyzed. From comparisons of the ASET with RSET, design suggestions were provided for evacuation slide spacing in undersea tunnels. In addition, the safe velocity modes of tunnel fire were proposed and the slow and high-risk areas for evacuation were identified. The conclusions are as follows:

(1) The low-velocity mode is a safer ventilation mode for downstream evacuations, such as 1.0 m/s used in this work. Turning off mechanical ventilation increases the environmental risk on both sides of the fire source and hampers the upstream rescue, so it is not recommended. Using a higher velocity increases the environmental risk downstream of the tunnel due to the destruction of smoke stratification.

(2) RSET decreases as the slide spacing D shortens; after $D < 60$ m, the decreasing rate is reduced from 39 s per additional slide to 11 s per additional slide. This leads to a decline in the cost-effectiveness of shortening evacuation time by adding slides.

(3) When using the low-velocity mode of 1.0 m/s, medium-velocity mode of 2.0 m/s, or high-velocity mode above 4.5 m/s in a tunnel fire, the respective reasonable evacuation spacings D should not be larger than 60 m, 50 m, and 30 m.

(4) The slow evacuating areas were always from 100 m to 300 m from the fire source and were independent of the slide spacing D. If $D \geq 70$ m, the downstream parts 200 m to 400 m from the fire source are always high-risk areas, independent of the ventilation velocity modes.

The ventilation velocity mode recommendation and the suggestions of reasonable evacuation spacings are based on the tunnel fire scenario assumed in this article. The conclusion provides design ideas for related projects, and it could be suitable for the evacuation designs of other tunnel projects by changing the parameters of tunnel model structure, traffic conditions, or fire scale. The limitations of the above work are that only a 50 MW fire of one typical undersea tunnel was simulated by a numerical method, for other scale fires such as a larger HRR, the corresponding ventilation velocity ranges, smoke hazards, and evacuation spacings may be different; these need to be reasonably calculated and extended to general cases, and the ventilation velocity was assumed to be the only factor affecting ASET. In addition, the coupled calculation method of fire and evacuation was not used; only the minimum margin of the safety evacuation methodologies was, which is ASET $\geq$ RSET. Different tunnel scenarios and fire scales with the influence of installation locations and the fan active time on ASET will be considered in future work. Experimental verification will be added as possible.

Author Contributions: Conceptualization, W.N. and C.C.; methodology, W.N. and C.C.; software, C.C.; validation, W.N. and C.C.; formal analysis, W.N. and C.C.; resources, W.N.; data curation, C.C.; writing—original draft preparation, C.C.; writing—review and editing, W.N.; visualization, W.N. and C.C.; supervision, W.N.; project administration, W.N.; funding acquisition, W.N. All authors have read and agreed to the published version of the manuscript.

Funding: This research was supported by the Beijing Natural Science Foundation (No. 9222009) and the Beijing Social Science Foundation (No. 20GLB025).

Institutional Review Board Statement: Not applicable.

Informed Consent Statement: Not applicable.

Data Availability Statement: Not applicable.

Conflicts of Interest: The authors declare no conflict of interest.

References

1. Huo, Y.; Gao, Y.; Chow, W.K. A study on ceiling jet characteristics in an inclined tunnel. *Tunn. Undergr. Space Technol.* **2015**, *50*, 32–46. [CrossRef]
2. Chen, H.; Chen, J.Z.; Tan, J.Q.; Zhang, J.Q. Intelligent Ventilation and Emergency Evacuation of Underwater Tunnel under fire Conditions. *Procedia Eng.* **2016**, *166*, 379–388. [CrossRef]
3. Chow, W.K.; Li, J.S. Case study: Vehicle fire in a cross-harbour tunnel in Hong Kong. *Tunn. Undergr. Space Technol.* **2001**, *16*, 23–30. [CrossRef]
4. Xu, Y.; Liao, S.M.; Liu, M.B. Simulation and Assessment of Fire Evacuation Modes for Long Underwater Vehicle Tunnels. *Fire Technol.* **2018**, *55*, 729–754. [CrossRef]
5. *ISO/TR 16738:2009(E)*; Fire-Safety Engineering—Technical Information on Methods for Evaluating Behavior and Movement of People. Technical Report for International Organization for Standardization; ISO: Geneva, Switzerland, 2009. Available online: https://www.iso.org/standard/42887.html (accessed on 10 March 2022).
6. Ji, J.; Guo, F.Y.; Gao, Z.H.; Zhu, J.P. Effects of ambient pressure on transport characteristics of thermal-driven smoke flow in a tunnel. *Int. J. Therm. Sci.* **2018**, *125*, 210–217. [CrossRef]
7. *JTG/T D70/2-20-2014*; Guidelines for Design of Ventilation of road Tunnel. Ministry of Communications of the People's Republic of China: Beijing, China, 2014.
8. Sturm, P.; Beyer, M.; Rafiei, M. On the Problem of Ventilation Control in Case of a Tunnel Fire Event. *Case Stud. Fire Saf.* **2017**, *7*, 36–43. [CrossRef]
9. Frey, S.; Lempp, M.; Zimmermann, C. Aspects of longitudinal airflow control in road tunnels. In Proceedings of the Virtual Conference Tunnel Safety and Ventilation 2020, Graz, Austria, 1–3 December 2020.
10. Technical Committee 3.3 Road Tunnel Operation. *Report 2011R02 Road tunnels: Operational Strategies for Emergency Ventilation*; PIARC: Paris, France, 2011; pp. 41–47. ISBN 2-84060-234-2.
11. Xue, H.; Chew, T.C.; Tay, K.L.; Cheng, Y.M. Control of Ventilation Airflow for Tunnel Fire Safety. *Combust. Sci. Technol.* **2000**, *152*, 179–196. [CrossRef]
12. Hu, X.Q. Numerical study of the effects of ventilation velocity on peak heat release rate and the confinement velocity in large tunnel fires. *Saf. Sci.* **2021**, *142*, 105359. [CrossRef]
13. Guo, Y.H.; Yuan, Z.Y.; Yuan, Y.P.; Cao, X.L.; Zhao, P. Numerical simulation of smoke stratification in tunnel fires under longitudinal velocities. *Undergr. Space* **2021**, *6*, 163–172. [CrossRef]
14. Liu, S. Study on Tunnel Fire Smoke Control and Transverse Passageway Interval under Traffic Congestion. In Proceedings of the IOP Conference Series: Earth and Environmental Science, Qingdao, China, 28–30 May 2021.
15. Jan, Š.; Lukáš, F.; Jiří, C.; Jiří, Z. Model-based airflow controller design for fire ventilation in road tunnels. *Tunn. Undergr. Space Technol.* **2016**, *60*, 121–134. [CrossRef]
16. Espinosa Antelo, I.; Fernández Martín, S.; Rey Llorente, I.d.; Alarcón Álvarez, E. Experiences on the specification of algorithms for fire and smoke control in road tunnels. In Proceedings of the 5th International Conference Tunnel Safety and Ventilation, Graz, Austria, 3–4 May 2010.
17. Yin, H.; Fang, Z. Evacuation Simulation of the Yangtze River-crossing Highway-metro Integrated Tunnel on Sanyang Road in Wuhan. In Proceedings of the 2019 9th International Conference on Fire Science and Fire Protection Engineering (ICFSFPE), Chengdu, China, 18–20 October 2019; pp. 1–6. [CrossRef]
18. Yuan, D.J.; Jin, H.; Chen, Z.C.; Liu, S.-N. Evacuation Experiment Study in Up and Down Escape Staircase of Underground Road. *Adv. Civ. Eng.* **2021**, *2021*, 10. [CrossRef]
19. Zhang, A.Y.; Deng, M.; Hu, Y.J. Reasonable Space Combination Evacuation of Contact Channel and Evacuation Slide for Two-Hole Shield Tunnel. *J. China Foreign Highw.* **2020**, *40*, 209–213. (In Chinese) [CrossRef]
20. Shen, Z.H.; Ma, J.; Qian, F.; Huang, Y.D. Simulation on Longitudinal Evacuation in Shield Tunnel under Fire. *Fire Sci. Technol.* **2019**, *38*, 92–94. (In Chinese)
21. Hui, Y.C.; Xiao, Y.M.; Huang, H.T. Study on the Design Parameters of Vertical Evacuation Channels for Highway Tunnel. *Refrig. Air Cond.* **2020**, *34*, 157–163. (In Chinese)
22. McGrattan, K.; Hostikka, S.; Floyd, J.; McDermott, R.; Vanella, M. Chapter 1Introduction. In *Fire Dynamics Simulator User's Guide*, 6th ed.; National Institute of Standards and Technology (NIST): Gaithersburg, MD, USA, 2020; pp. 3–9. [CrossRef]
23. Tang, F.; Deng, L.; Meng, N.; Mcnamee, M.; Van Hees, P.; Hu, L.H. Critical longitudinal ventilation velocity for smoke control in a tunnel induced by two nearby fires of various distances: Experiments and a revisited model. *Tunn. Undergr. Space Technol.* **2020**, *105*, 103559. [CrossRef]
24. Chow, W.K.; Wong, K.; Chung, W. Longitudinal ventilation for smoke control in a tilted tunnel by scale modeling. *Tunn. Undergr. Space Technol.* **2010**, *25*, 122–128. [CrossRef]
25. *ISO 13943:2017(E)*; Fire Safety—Vocabulary. Technical report for International Organization for Standardization; ISO: Geneva, Switzerland, 2017. Available online: https://www.iso.org/standard/63321.html (accessed on 10 March 2022).
26. *ISO 13571:2012(E)*; Life-Threatening Components of Fire—Guidelines for the Estimation of Time to Compromised Tenability in Fires. Technical report for International Organization for Standardization; ISO: Geneva, Switzerland, 2012. Available online: https://www.iso.org/standard/56172.html (accessed on 10 March 2022).

27. Wei, W. Study on the Numerical Simulation and Safety Evacuation in Long Highway Tunnel Fire. Master's Thesis, Chang'an University, Xi'an, China, 2008. (In Chinese).
28. Purser, D.A.; McAllister, J.L. Assessment of Hazards to Occupants from Smoke, Toxic Gases, and Heat. In *SFPE Handbook of Fire Protection Engineering*; Hurley, M.J., Gottuk, D., Hall, J.R., Harada, K., Kuligowski, E., Puchovsky, M., Torero, J., Watts, J.M., Wieczorek, C., Eds.; Springer: New York, NY, USA, 2016; pp. 2308–2428.
29. Guo, X.X.; Pan, X.X.; Zhang, L.J.; Wang, Z.L.; Hua, M.; Jiang, J.C. Comparative study on ventilation and smoke extraction systems of different super-long river-crossing subway tunnels under fire scenarios. *Tunn. Undergr. Space Technol.* **2021**, *113*, 103849. [CrossRef]
30. Król, A.; Król, M. Numerical investigation on fire accident and evacuation in a urban tunnel for different traffic conditions. *Tunn. Undergr. Space Technol.* **2021**, *109*, 103751. [CrossRef]
31. Thunderhead Engineering. Chapter 6. Steering Mode. In *Pathfinder Technical Reference Manual*; Thunderhead Engineering Consultants, Inc: Manhattan, NY, USA, 2021; pp. 29–43. Available online: https://www.thunderheadeng.com (accessed on 10 March 2022).
32. Song, X.Y.; Pan, Y.; Jiang, J.C.; Wu, F.; Ding, Y.J. Numerical investigation on the evacuation of passengers in metro train fire. *Procedia Eng.* **2018**, *211*, 644–650. [CrossRef]
33. Zhang, G.W.; Huang, D.; Zhu, G.Q.; Yuan, G.L. Probabilistic model for safe evacuation under the effect of uncertain factors in fire. *Saf. Sci.* **2017**, *93*, 222–229. [CrossRef]
34. Seike, M.; Kawabata, N.; Hasegawa, M. Evacuation speed in full-scale darkened tunnel filled with smoke. *Fire Saf. J.* **2017**, *91*, 901–907. [CrossRef]
35. Technical Committee 5 Road Tunnels. *Report 05.05.B Fire and Smoke Control in Road Tunnels*; PIARC: Paris, France, 1999; pp. 31–33. ISBN 2-84060-064-1.
36. Xie, B.C.; Zhang, S.Q.; Xu, Z.S.; He, L.; Xi, B.H.; Wang, M.Q. Experimental study on vertical evacuation capacity of evacuation slide in road shield tunnel. *Tunn. Undergr. Space Technol.* **2020**, *97*, 103250. [CrossRef]
37. Zhang, Y.C.; Zhou, A.X.; Xiang, Y.; He, C.; Jiao, Q.Z.; Wan, B.; Xie, W. Evacuation experiments in vertical exit passages in an underwater road shield tunnel. *Phys. A: Stat. Mech. Its Appl.* **2018**, *512*, 1140–1151. [CrossRef]
38. Cao, Y. Research on Experiment for Vertical Evacuation of Underwater Highway Tunnel with Large Cross-Section. Master's Thesis, Southwest Jiaotong University, Chengdu, China, 2016. (In Chinese).
39. Ren, R.; Zhou, H.; Hu, Z.; He, S.Y.; Wang, X.L. Statistical analysis of fire accidents in Chinese highway tunnels 2000–2016. *Tunn. Undergr. Space Technol.* **2019**, *83*, 452–460. [CrossRef]
40. Haack, A. Current safety issues in traffic tunnels. *Tunn. Undergr. Space Technol.* **2002**, *17*, 117–127. [CrossRef]
41. Technical Committee 3.3 Road Tunnel Operation. *Report 2017R01 Design Fire Characteristics for Road Tunnels*; PIARC: Paris, France, 2017; pp. 5–7. ISBN 978-2-84060-471-6.
42. Tewarson, A. Section 3, Chapter 4, Generation of Heat and Chemical Compounds in Fires. In *The SFPE Handbook of Fire Protection Engineering*, 2nd ed.; DiNenno, P.J., Ed.; National Fire Protection Association: Quincy, MA, USA, 1995; pp. 3-53–3-124.
43. Hu, L.H.; Fong, N.K.; Yang, L.Z.; Chow, W.K.; Li, Y.Z.; Huo, R. Modeling fire-induced smoke spread and carbon monoxide transportation in a long channel: Fire dynamics simulator comparisons with measured data. *J. Hazard. Mater.* **2007**, *140*, 293–298. [CrossRef]
44. Gehandler, J.; Eymann, L.; Regeffe, M. Limit-based fire hazard model for evaluating tunnel life safety. *Fire Technol.* **2015**, *51*, 585–614. [CrossRef]

fire

Article

Study of the Heat Exhaust Coefficient of Lateral Smoke Exhaust in Tunnel Fires: The Effect of Tunnel Width and Transverse Position of the Fire Source

Qiulin Liu , Zhisheng Xu, Weikun Xu, Sylvain Marcial Sakepa Tagne, Haowen Tao, Jiaming Zhao and Houlin Ying *

School of Civil Engineering, Central South University (CSU), Changsha 410075, China
* Correspondence: evelynlin@csu.edu.cn; Tel.: +86-731-8265-5177

Abstract: The tunnel width and the transverse fire's position both impact the heat exhaust coefficient, which is a critical component of the lateral smoke exhaust in tunnel fires. In this research, the tunnel width and the transverse location of the fire source are varied to analyze the heat exhaust coefficient of lateral smoke exhaust. When tunnel width increases, there is a noticeable decrease in the longitudinal temperature of the entrained air and smoke layer in the fire plume. Furthermore, the heat exhaust coefficients are reduced. An increase in the distance between the exhaust vent and the fire source causes an increase in the proportion of hot smoke in the smoke exhaust mass flow, which increases the heat exhaust coefficient. A calculated heat exhaust coefficient was developed using the fire source's location and the tunnel's width as inputs, which agrees well with the simulation results. This method can predict the heat exhaust coefficient of the lateral smoke exhaust in tunnel fires. The findings of this study provide insight into how the tunnel width and the location of a transverse fire influence the heat exhaust coefficient.

Keywords: tunnel fire; heat exhaust coefficient; tunnel width; transverse fire location; lateral smoke exhaust

Citation: Liu, Q.; Xu, Z.; Xu, W.; Tagne, S.M.S.; Tao, H.; Zhao, J.; Ying, H. Study of the Heat Exhaust Coefficient of Lateral Smoke Exhaust in Tunnel Fires: The Effect of Tunnel Width and Transverse Position of the Fire Source. *Fire* **2022**, *5*, 167. https://doi.org/10.3390/fire5050167

Academic Editors: Dahai Qi, Tiago Miguel Ferreira and Fei Tang

Received: 4 September 2022
Accepted: 10 October 2022
Published: 15 October 2022

1. Introduction

The development of construction technology has been flourishing, and the number of tunnels with large diameter increases continually [1]. However, the studies on fire safety in tunnels with large diameters are still insufficient [2]. Since the heat dissipation is restricted and the hot smoke spreads longitudinally along the tunnel during tunnel fires [3,4], using a reasonable ventilation system to prevent the spread of smoke and harmful gases is crucial [5,6]. The lateral smoke exhaust has been adopted in various tunnels [7]. Nevertheless, the change in the tunnel width [8,9] and the transverse position of the fire source [10,11] influence the heat exhaust coefficient.

According to Chen et al.'s study [12], the maximum temperature rise of the fire and the attenuation coefficient of smoke flow influenced the heat lost by the smoke exhaust. However, the tunnel width influenced the maximum temperature rise and the attenuation factor of smoke close to the fire source. [13]. Additionally, this investigation did not consider the amount of smoke in the expelled mass flow. The amount of smoke in the exhaust mass flow caused by air entrainment at the exhaust vent is affected by the plug-holing phenomena. The plug-holing phenomenon would also be experienced by the lateral smoke exhaust, according to Yang et al. [14]. When plug-holing occurs, a great deal of cold air is entrained, which reduces the efficiency of smoke extraction. Jiang et al. [15] calculated the smoke back-layering length in their study. They redefined the confinement velocity based on the smoke temperature distribution below the tunnel ceiling. Zhong et al. [16] proposed the concept of a virtual smoke vent based on the results of different smoke exhaust velocities, heat release rates and smoke layer heights in their study of the occurrence of plug-holing penetration in the lateral smoke exhaust. A critical Froude number for predicting lateral smoke exhaust plug-holing was established and a corresponding critical Froude number of 0.5 was given for the occurrence of plug-holing. Zhu et al. [17] proposed a modified Froude number for determining the plug-holing phenomenon through a tunnel fire under the action of lateral smoke exhaust. The critical Froude number ranges from approximately 1.5 to 1.75. Our previous study [18] considered the difference between lateral smoke exhaust and ceiling smoke exhaust. We modified the Froude number to determine the plug-holing of lateral smoke exhaust to establish a new modified Froude number. The results show that when the new modified Froude number is greater than 2.5, the lateral smoke exhaust experiences plug-holing.

However, previous research has concentrated on determining plug-holing in the lateral smoke exhaust. There is a lack of research into the effect of different tunnel widths and the lateral position of the fire source on the effectiveness of lateral smoke exhaust. Additionally, the direction of extraction is different for the ceiling smoke exhaust and the lateral smoke exhaust, especially in large-diameter tunnels. The cold air is encased by the smoke layer and exhaust vent working together. Meanwhile, there is a lack of correlation between the tunnel width, the transverse fire location, and the heat exhaust efficiency. Therefore, by large eddy simulation (LES) and theoretical analysis, the change of the influencing factors heat exhaust coefficient in tunnel fires with various tunnel widths and transverse fire locations was studied.

2. Theoretical Analysis

The heat exhaust coefficient determines the quantity of heat expelled [19], which is the ratio of heat exhausted to heat dispersed from the fire source to one side of the tunnel.

$$\eta = \frac{\dot{Q}_{es}}{C_x \dot{Q}_c} = \frac{c_p \dot{m}_s \Delta T_r}{C_x \dot{Q}_c} = \frac{c_p \varphi_s \dot{m}_{es} \Delta T_r}{C_x \dot{Q}_c} \tag{1}$$

where:
$\dot{Q}_{es}$: heat carried by the smoke discharged from exhaust vent (kW);
C_x: correction factor;
$\dot{Q}_c$: convective heat release rate (kW);
c_p: thermal capacity of air (kJ/(kg·K));
$\dot{m}_s$: mass flow of the hot smoke exhausted by the exhaust vent (kg/s);
φ_s: proportion of smoke in the exhausted mass flow;
$\dot{m}_{es}$ mass flow of the exhausted gas (kg/s);
ΔT_r: temperature increase in the smoke layer at the exhaust vent (K).

According to the study of Ingason et al. [20], radiative heat constitutes between 20 and 40 percent of total heat release, as a result, $\dot{Q}_c$ can be described as:

$$\dot{Q}_c = 0.7\dot{Q} \tag{2}$$

where:
$\dot{Q}$: total heat release rate of the fire source (kW).

For a particular fire scenario, $\dot{Q}_c$ and $\dot{m}_{es}$ are almost constant. According to Equation (1), it can be inferred that the heat exhaust coefficient mainly relates to φ_s and ΔT_r.

As shown in Figure 1, the flow field beside the exhaust vent becomes a disturbance, and hot smoke is mixed with cold air more intensely during the exhausting process. $\dot{m}_{a1}$ is a representation of the mass flow of the cold air directly discharged by the exhaust vent, $\dot{m}_{a2}$ is a representation of the mass flow of the cold air entrained by the smoke layer.

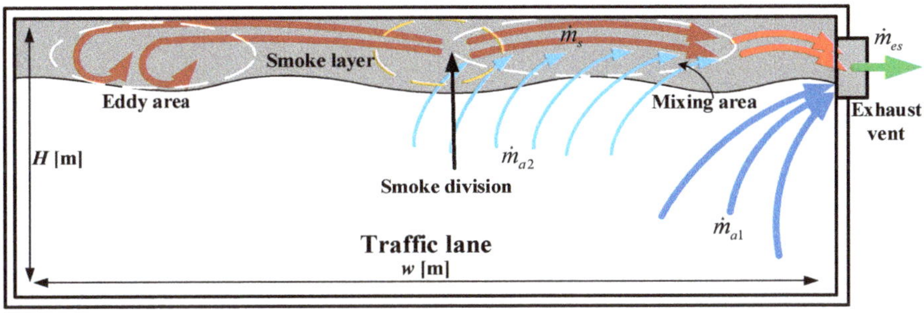

Figure 1. Schematic diagram of the flow field beside the exhaust vent.

According to the principle of conservation of mass, it can be obtained that:

$$\dot{m}_{es} = \dot{m}_s + \dot{m}_{a1} + \dot{m}_{a2} \tag{3}$$

φ_{a1} is the proportion of $\dot{m}_{a1}$ in the exhausted mass flow:

$$\varphi_{a1} = \frac{\dot{m}_{a1}}{\dot{m}_{es}} \tag{4}$$

In the early research, Hinkley [21] analyzed the plug-holing phenomenon, which was prompted by natural smoke extraction. Hinkley proposed a dimensionless number Fr to show the plug-holing that took place, which can be expressed as Equation (5).

$$Fr = \frac{u_v A}{(g\Delta T/T_0)^{1/2} d^{5/2}} \tag{5}$$

where:
g: gravitational acceleration (m/s^2);
ΔT: temperature rise of the smoke layer (K);
d: thickness of the smoke layer (m);
u_v: flow velocity at the exhaust vent (m/s);
A: area of the exhaust vent (m^2).

The study of Ji et al. [22] has shown that the air entrainment in the exhausted mass flow could increase to 48%, where no apparent plug-holing phenomenon is observed. The Fr number might be intimately connected to the exhaust mass flow's air entrainment. Furthermore, according to Cong et al. [23], the Fr value may be used to calculate the air entrainment at the exhaust vent. So, the values of Fr are related to the proportion of $\dot{m}_{a1}$ in $\dot{m}_{es}$, i.e.,

$$\varphi_{a1} = C_f Fr = \frac{C_f u_v A}{(g\Delta T/T_0)^{1/2} d^{5/2}} \tag{6}$$

where:
C_f: correction factor.

As shown in Figure 1, the division phenomenon of the smoke layer occurs under the dual function of the eddy area and the smoke exhaust vent, and cold air $\dot{m}_{a2}$ is sucked into the smoke layer, φ_{a2} is a representation of the ratio of $\dot{m}_{a2}$ to the mass flow of the exhausted gas ($\dot{m}_{es}$):

$$\varphi_{a2} = \frac{\dot{m}_{a2}}{\dot{m}_{es}} \tag{7}$$

According to Equation (3), the percentage of smoke in the exhausted mass flow (φ_s), can be expressed as:

$$\varphi_s = 1 - \varphi_{a1} - \varphi_{a2} = 1 - \frac{C_f u_v A}{(g\Delta T/T_0)^{1/2} d^{5/2}} - \frac{\dot{m}_{a2}}{\dot{m}_{es}} \tag{8}$$

In the FDS simulation, ΔT can be measured by placing thermocouples; u_v can be measured by placing a "Velocity Device" at the exhaust vent; d can be measured by setting up a "Layer Zoning Device"; $\dot{m}_{es}$ can be measured by placing a "Mass Flow Device" at the exhaust vent; $\dot{m}_s$ can be measured at the cross-section of the tunnel with the "Mass Flow Device". The parameters measured in different scenarios are substituted into Equation (8) and then the remaining unknown parameters are obtained by associating Equation (8) with different scenarios.

3. Numerical Modeling

FDS has been extensively applied in fire and smoke control research and has been widely verified for its reliability. The Large Eddy Simulation (LES), which uses a filter function to separate the large eddy from the small eddy, is used in FDS to solve the Reynolds stress. Numerical computations are used to unravel the large eddy, while Sub-Grid Models (SGM) are used to unravel the small eddy.

3.1. Model Design

As illustrated in Figure 2, the tunnel model used in this study is a full-size tunnel model of 200 m long, 7 m high and 10/15/20/25 m wide. The wall material of the tunnel in this paper is concrete, with a friction factor set to 0.022. In addition to a Prandtl number of FDS default value of 0.5. The Stanton number was calculated from the above parameters to be 0.017. The fire source is situated at varied transverse distances from the tunnel side wall and 50 m from the exit at the left end of the tunnel. In addition, the fire source is 0.5 m from the ground. The lateral smoke exhaust vent of 4 m × 1.2 m is positioned 80 m from the exit at the left end of the tunnel.

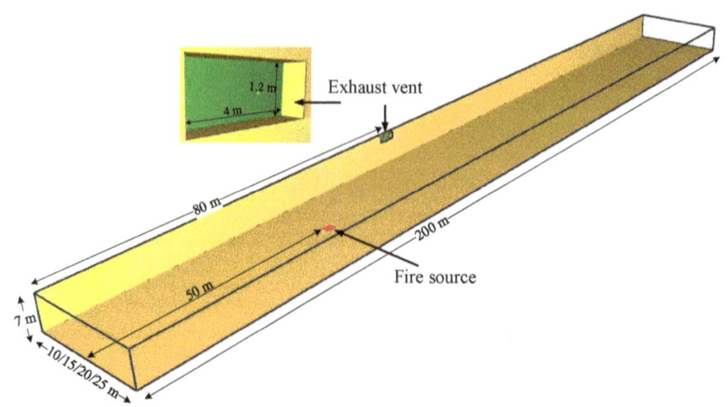

Figure 2. Model configuration.

Numerous studies [24–26] have found the most common fire scenarios to be 3–5 MW for car fires and 10–20 MW for truck fires in road tunnels. So, the heat release rate is set at 12 MW in this paper. There was a 293 K ambient temperature and a 101 kPa ambient pressure set. The border conditions of the tunnel's two portals were designated as "OPEN". A 400 s simulation was conducted. The grid size we chose is 0.25 m to simulate the tunnel fire.

As demonstrated in Figure 3, the thermocouples positioned 0.2 m below the ceiling centerline and spaced 1 m were developed to detect the temperature distribution at the ceiling. Temperatures of the smoke were measured using thermocouples positioned vertically, placed 15 m from the fire source, and spaced 0.2 m. The "Layer Zoning Device" in FDS was used to gauge the smoke layer where the exhaust vent is located, which can measure the thickness of the smoke layer. A "Mass Flow Device" and a "Heat Flow Device" was mounted inside the smoke layer to measure the amount of heat transported by the smoke layer per unit mass at the site of the exhaust vent. Meanwhile, the exhaust vent's heat and mass flow were detected by the function of the "Mass Flow Device" and "Heat Flow Device" placed in front of the exhaust vent. A velocity measurement was made at the exhaust vent using the "Velocity Device" in FDS.

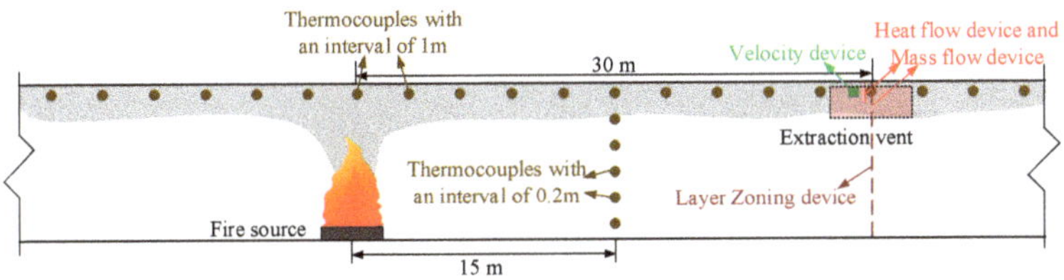

Figure 3. Probe arrangement.

A set of temperature and velocity slices were set up at the smoke vent and the lateral position of the fire source to obtain the temperature and velocity flow fields at the fire source and smoke vent. The slices were processed with Tecplot software to obtain the temperature and velocity flow fields. The average value of the stable section from 300 s to 400 s was extracted from the simulation results for discussion.

3.2. Simulated Conditions

Table 1 shows the parameters under every 32 scenarios in this research. In this study, two mass flow rates ($\dot{m}_{es}$) of 0 kg/s (no smoke exhaust) and 12 kg/s (smoke exhaust) were determined for the smoke exhaust vent. The width of the tunnel increased from 10 m adding 5 m in sequence up to 25 m. The fire sources were in different transverse locations of the tunnel.

Table 1. Summary of the scenarios.

NO.	HRR (MW)	Mass Flow Rate of the Exhaust Vent (kg/s)	Width of the Tunnel (m)	Transverse Distance between Fire Source and Side Wall (m)
1–3	12	0	10	2.5, 5, 7.5
4–6			15	2.5, 7.5, 12.5
7–11			20	2.5, 7.5, 10, 12.5, 17.5
12–16			25	2.5,7.5, 12.5, 17.5, 22.5
17–19		12	10	2.5, 5, 7.5
20–22			15	2.5, 7.5, 12.5
23–27			20	2.5, 7.5, 10, 12.5, 17.5
28–32			25	2.5,7.5, 12.5, 17.5, 22.5

We calculated the maximum temperature rise below the tunnel ceiling (tunnel width: 25 m; transverse distance between fire source and side wall: 2/3/4/5/6 m) and compared it with the findings of the model-scale test of Ji et al. [27] (Equation (9)), as shown in Figure 4. The results of the numerical simulations are in general agreement with the results of the model-scale test. However, there are still some differences between the two results. The reasons for the deviation between the two results are complex. Firstly, due to the tightness of the assumptions made in the numerical simulation model, there is some influence on the maximum temperature rise below the tunnel ceiling. In their experiments, Ji et al. did not specify any parameters for the tunnel wall material. As a result, the materials used in Ji et al.'s model-scale test have different specific heat values, densities, and thermal conductivities than the concrete described in this work. Which may also alter the maximum temperature rise under the tunnel ceiling. However, the results of the numerical simulations in this paper agree with the data from the previous model-scale test. The numerical simulations can therefore be used to investigate the fire source's smoke flow at different widths and lateral positions.

$$\Delta T_{\max,D} = 17.9\frac{\dot{Q}^{2/3}}{H^{5/3}}(1.096e^{-14.078D/(w/2)}+1) \tag{9}$$

where:
$\Delta T_{\max,D}$: maximum temperature rise at a distance D from the side wall (K);
H: tunnel height (m);
D: distance from the side wall (m);
w: tunnel width (m).

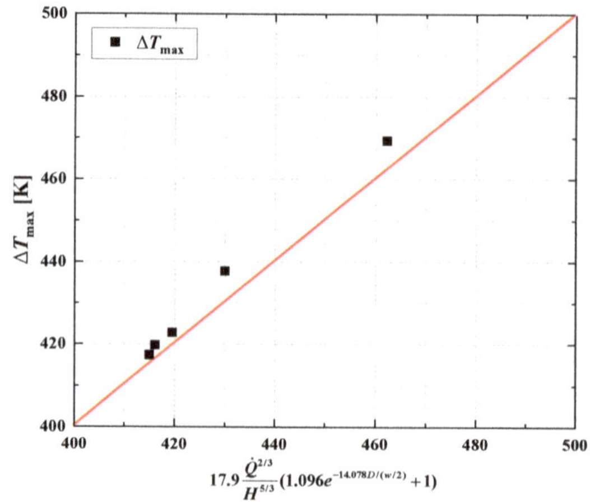

Figure 4. Comparison of the maximum temperature rise at different locations with Ji's model (Equation (9)).

4. Results and Discussion

The variation of the factors influencing the heat exhaust coefficient was studied by varying two variables: the fire source's transverse position and the tunnel's width. A model for predicting the heat exhaust coefficient in tunnel fires was proposed.

4.1. *Effect of Tunnel Width on Fire Plume without Smoke Exhaust*

Since tunnel width increases with tunnel diameter, there is a change in both the transverse spread distance of smoke and the cold air entrained by the fire plume. Figure 5 demonstrates the development of fire plumes in tunnels with different widths. When the tunnel width is 10 m and 15 m, the fire plume's air entrainment in the tunnel's transverse direction is less than when the tunnel is wider. Most of the air between the tunnel side wall and the fire source gets heated significantly when the tunnel width is 10 m. Air entrainment by the fire plume behaves similarly as tunnel width increases, decreasing smoke temperatures. In the diffuse stage of smoke flow, when the smoke flow diffuses to the tunnel sidewall, an anti-buoyancy wall jet is formed by the smoke flow moving vertically downward alongside the sidewall, resulting in the backflow of some smoke. As the width of the tunnel increases, the greater the resistance in the process of diffusion of smoke to the tunnel sidewalls, resulting in a gradual weakening of the smoke's horizontal inertia force. Therefore, the wider the tunnel, the weaker the anti-buoyancy wall jet, and its vertical length decreases significantly. The smoke layer's thickness also tends to thin as tunnel width increases. The wider the tunnel, the thinner the smoke layer will be under the same smoke production volume.

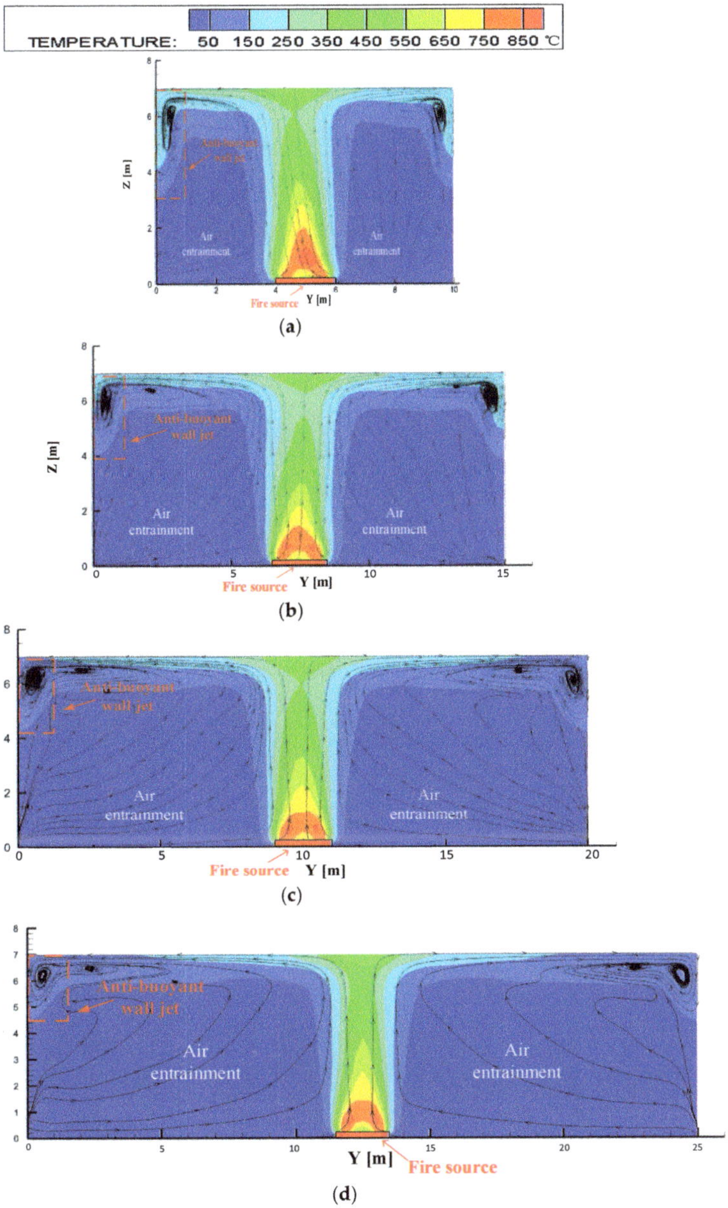

Figure 5. Temperature and velocity flow distribution around the fire source for different tunnel widths: (**a**) 10 m; (**b**) 15 m; (**c**) 20 m; (**d**) 25 m.

4.2. *Effect of Fire Source Transverse Location on Heat Exhaust Coefficient*

As the smoke from the fire source hits the side walls under the effect of the transverse inertial force, as depicted in Figure 6, an eddy area forms under the ceiling. The eddy area under the ceiling on the side farther from the fire source becomes larger. The size of the vortex area affects the heat exhaust coefficient. The eddy area also spreads forward during the longitudinal spread of smoke, affecting lateral smoke exhaust. As depicted in Figure 7, when the fire source is situated at distances of 2.5 m and 7.5 m, there is a large range of strong eddy areas under the ceiling far from the exhaust vent, resulting in the division of hot smoke. A mass of cold air enters the smoke layer and mixes with the hot smoke, ultimately decreasing the heat exhaust coefficient. When the fire source is positioned at the center of 12.5 m, then the range of the strong eddy area is reduced. Although the hot smoke diversion phenomenon still occurs, the position of smoke diversion is farther away from the exhaust vent. The extraction exhausts more hot smoke, improving the heat exhaust coefficient. If the fire source is positioned at 17.5 m and 22.5 m, there is no obvious eddy area and smoke diversion phenomenon in the smoke layer. Most of the hot smoke moves toward the exhaust vent, effectively reducing the mixing of cold air and significantly improving the heat exhaust coefficient.

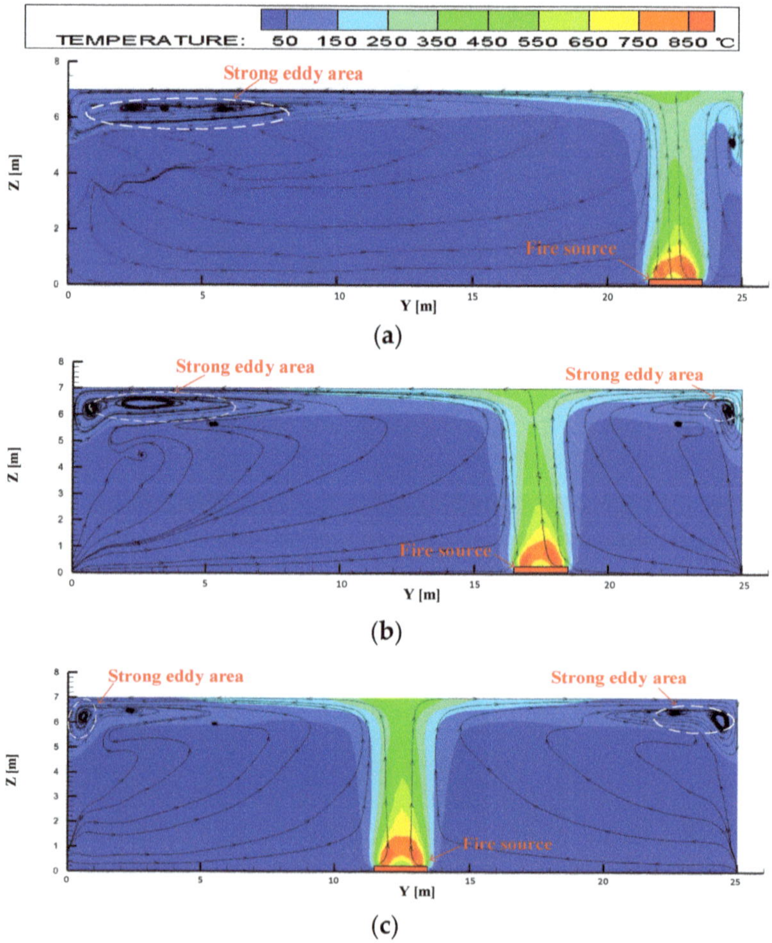

Figure 6. *Cont.*

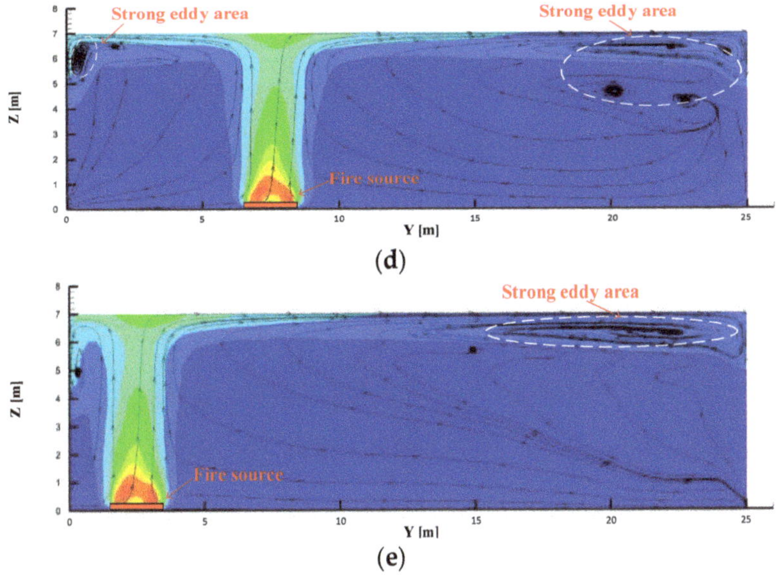

Figure 6. Temperature and velocity flow distribution around the fire source in the 25 m wide tunnel for different distances from the side walls: (**a**) 2.5 m; (**b**) 7.5 m; (**c**) 12.5 m; (**d**) 17.5 m; (**e**) 22.5 m.

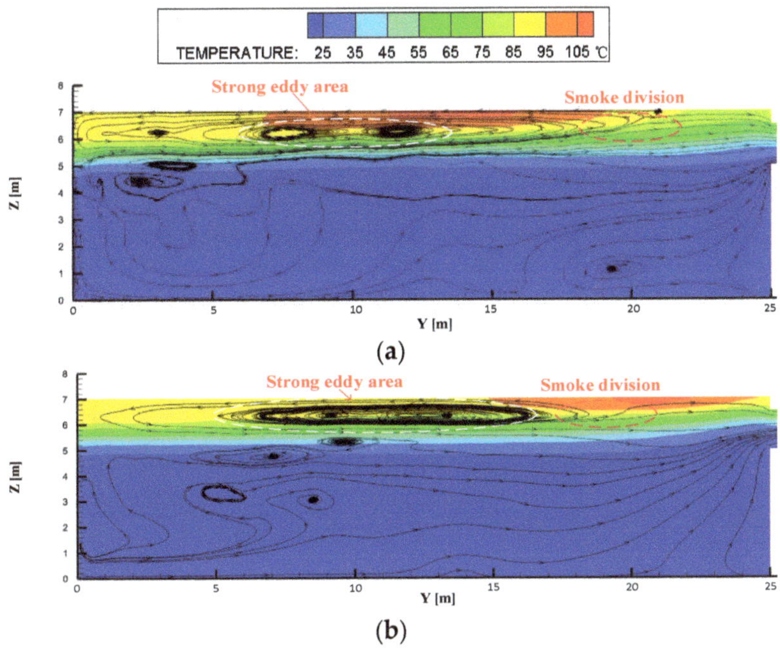

Figure 7. *Cont.*

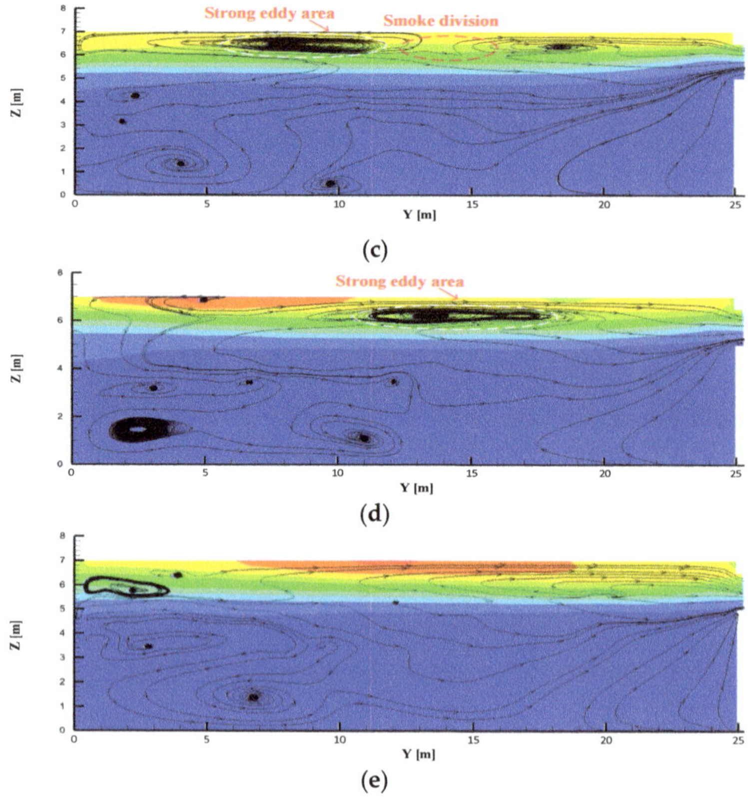

Figure 7. Temperature and velocity flow distribution around the exhaust vent in the 25 m wide tunnel for different distances from the side walls: (**a**) 2.5 m; (**b**) 7.5 m; (**c**) 12.5 m; (**d**) 17.5 m; (**e**) 22.5 m.

It can be deferred that the ratio of the transverse fire source distance from the exhaust vent (L) to the tunnel width (w) will affect the φ_{a2} by observing Figures 6 and 7. So, φ_{a2} can be described as:

$$\varphi_{a2} = f(\frac{L}{w}) \tag{10}$$

Figures 8 and 9 show the correlation between the proportion of hot smoke and cool air in the smoke exhausted by the lateral exhaust vent in the tunnel that is 20 m and 25 m wide. When there is an increase in the transverse distance between the lateral exhaust vent and the fire source, the proportion of hot smoke increases while the proportion of cold air drops.

In the 20 m wide tunnel, the following formula can be used to represent the percentage of cold air in the exhausted mass flow of the lateral exhaust vent:

$$\varphi_{a1} + \varphi_{a2} = -0.088 \ln(\frac{L}{w}) + 0.5156 \tag{11}$$

In the 25 m wide tunnel, the following formula can be used to represent the percentage of cold air in the exhausted mass flow of the lateral exhaust vent:

$$\varphi_{a1} + \varphi_{a2} = -0.088 \ln(\frac{L}{w}) + 0.5361 \tag{12}$$

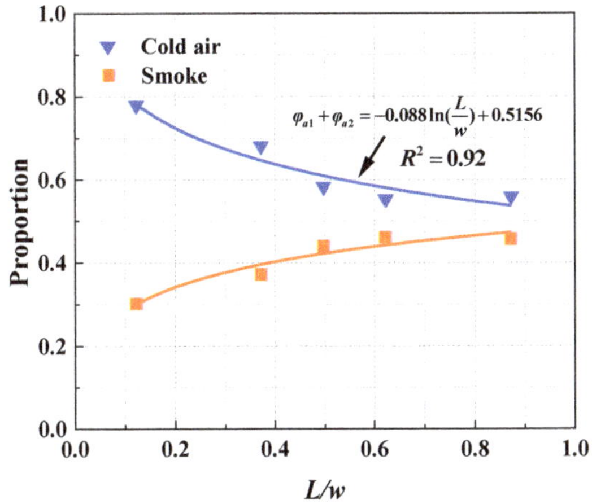

Figure 8. The proportion of hot smoke and cold air in the exhaust gas in a 20 m wide tunnel.

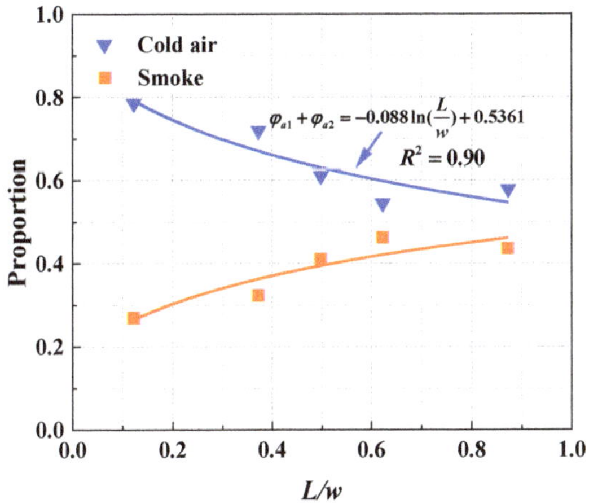

Figure 9. The proportion of hot smoke and cold air in the exhaust gas in a 25 m wide tunnel.

The value of Fr is 2.04 for a tunnel with a width of 20 m and 2.41 for a tunnel with a width of 25 m. When Equation (6) is combined with Equations (10)–(12), the following results are produced:

$$\begin{cases} 2.04C_f + f(\frac{L}{w}) = -0.088\ln(\frac{L}{w}) + 0.5156 \\ 2.41C_f + f(\frac{L}{w}) = -0.088\ln(\frac{L}{w}) + 0.5361 \end{cases} \quad (13)$$

Then, the value of C_f can be obtained as 0.056. So φ_{a1} can be expressed as:

$$\varphi_{a1} = 0.056Fr \quad (14)$$

Substituting Equation (14) into Equation (12), φ_{a2} can be expressed as:

$$\varphi_{a2} = -0.088\ln(\frac{L}{w}) + 0.40 \quad (15)$$

Substituting Equations (14) and (15) into Equation (8), φ_s can be expressed as:

$$\varphi_s = 0.6 + 0.088\ln(\frac{L}{w}) - 0.056Fr \tag{16}$$

The amount of hot smoke in the expelled mass flow grows as the transverse distance between the lateral exhaust vent and the fire source increases, ultimately leading to a larger heat exhaust coefficient.

4.3. Verification of the Effect of Lateral Extraction Efficiency

The value of C_x (Equation (1)) can be obtained to be 0.44, combining the simulation results obtained from scenarios 1~32. Then the heat exhaust coefficient can be described as:

$$\begin{cases} \eta = \frac{\Delta T_r \varphi_s \dot{m}_{es}}{0.44\dot{Q}_c} \\ \varphi_s = 0.6 + 0.088\ln(\frac{L}{w}) - 0.056Fr \end{cases} \tag{17}$$

As shown in Figure 10, Equation (17) matches the simulation results quite well. It is clear in Figure 10 that the heat exhaust coefficient increases with the transverse interval from the fire to the exhaust vent, while it declines with the growth of the tunnel width.

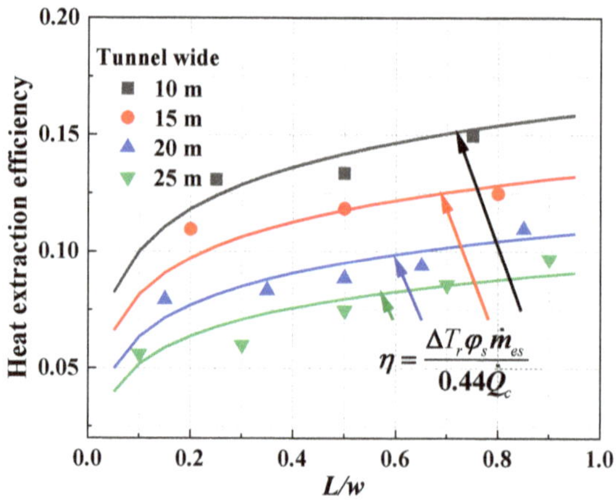

Figure 10. Comparison of Equation (17) with numerical simulation results.

5. Conclusions

This paper used a series of LES methods to conduct a simulation. The heat exhaust coefficient associated with tunnel fire lateral smoke exhaust has been investigated. A correlation for predicting the heat exhaust coefficient has been proposed. The main conclusions are:

(1) There will be an increase in the amount of hot smoke in the exhaust mass flow if the transverse distance between the exhaust vent and the fire source increases, increasing the heat exhaust coefficient in tunnel fires.

(2) The more expansive the tunnel, the weaker the anti-buoyancy wall jet, and its vertical length decreases significantly. In addition, as the tunnel width grows, the smoke layer thickness becomes gradually thinner.

(3) A calculation model was developed to compute the heat exhaust coefficient (Figure 10, Equation (17)) taking into account the lateral position of the fire source and the width of the tunnel.

Author Contributions: Methodology, Z.X.; software, W.X.; validation, H.T.; formal analysis, Q.L.; investigation, Q.L.; resources, H.Y.; data curation, J.Z.; writing—original draft preparation, Q.L.; writing—review and editing, S.M.S.T. All authors have read and agreed to the published version of the manuscript.

Funding: The Fundamental Research Funds for the Central Universities of Central South University (No. 2021zzts235) provided funding for this project.

Institutional Review Board Statement: Not applicable.

Informed Consent Statement: Not applicable.

Data Availability Statement: Not applicable.

Acknowledgments: We acknowledge the High Performance Computing Center of Central South University for its support. This work was supported by the Natural Science Foundation of Hunan Province of China (Grant No. 2020JJ3046) and the Fundamental Research Funds for the Central Universities of Central South University (Grant No. 2021zzts235).

Conflicts of Interest: The authors declare no conflict of interest.

References

1. Zhang, F.; Gao, Y.; Wu, Y.; Wang, Z. Face stability analysis of large-diameter slurry shield-driven tunnels with linearly increasing undrained strength. *Tunn. Undergr. Space Technol.* **2018**, *78*, 178–187. [CrossRef]
2. Xu, Z.S.; Liu, Q.L.; He, L.; Tao, H.W.; Zhao, J.M.; Chen, H.G.; Li, L.J.; Fan, C.G. Study on the heat exhaust coefficient and smoke flow characteristics under lateral smoke exhaust in tunnel fires. *Fire Mater.* **2019**, *43*, 857–867. [CrossRef]
3. Yao, Y.Z.; Li, Y.Z.; Ingason, H.; Cheng, X.D. Numerical study on overall smoke control using naturally ventilated shafts during fires in a road tunnel. *Int. J. Therm. Sci.* **2019**, *140*, 491–504. [CrossRef]
4. Yao, Y.Z.; He, K.; Peng, M.; Shi, L.; Cheng, X.D.; Zhang, H.P. Maximum gas temperature rise beneath the ceiling in a portals-sealed tunnel fire. *Tunn. Undergr. Space Technol.* **2018**, *80*, 10–15. [CrossRef]
5. Beard, A.N. Fire safety in tunnels. *Fire Saf. J.* **2009**, *44*, 276–278. [CrossRef]
6. Fan, C.G.; Ji, J.; Sun, J.H. Influence of longitudinal fire location on smoke characteristics under the tunnel ceiling. *Fire Mater.* **2015**, *39*, 72–84. [CrossRef]
7. Zhang, X.; Xu, Z.S.; Ni, T.X.; Peng, J.Z.; Zeng, J.W.; Ran, Q.B. Investigation on smoke temperature distribution in a double-deck tunnel fire with longitudinal ventilation and lateral smoke extraction. *Case Stud. Therm. Eng.* **2019**, *13*, 100375. [CrossRef]
8. Xu, Z.S.; Zhao, J.M.; Liu, Q.L.; Chen, H.G.; Liu, Y.H.; Geng, Z.Y.; He, L. Experimental investigation on smoke spread characteristics and smoke layer height in tunnels. *Fire Mater.* **2019**, *43*, 303–309. [CrossRef]
9. Wang, Y.F.; Sun, X.F.; Liu, S.; Yan, P.N.; Qin, T.; Zhang, B. Simulation of back-layering length in tunnel fire with vertical shafts. *Appl. Therm. Eng.* **2016**, *109*, 344–350. [CrossRef]
10. Fan, C.G.; Chen, J.; Mao, Z.L.; Zhou, Y.; Mao, S.H. A numerical study on the effects of naturally ventilated shaft and fire locations in urban tunnels. *Fire Mater.* **2019**, *43*, 949–960. [CrossRef]
11. Fan, C.G.; Chen, J.; Zhou, Y.; Liu, X.P. Effects of fire location on the capacity of smoke exhaust from natural ventilation shafts in urban tunnels. *Fire Mater.* **2018**, *42*, 974–984. [CrossRef]
12. Chen, L.F.; Hu, L.H.; Zhang, X.L.; Zhang, X.Z.; Zhang, X.C.; Yang, L.Z. Thermal buoyant smoke back-layering flow length in a longitudinal ventilated tunnel with ceiling extraction at difference distance from heat source. *Appl. Therm. Eng.* **2015**, *78*, 129–135. [CrossRef]
13. Ingason, H.; Li, Y.Z. Model scale tunnel fire tests with longitudinal ventilation. *Fire Saf. J.* **2010**, *45*, 317–384. [CrossRef]
14. Yang, J.; Pan, X.H.; Hua, M. Study on the Flow-dynamic Characterization of the Lateral Mechanical extraction system in tunnel fire. *Procedia Eng.* **2018**, *211*, 888–896. [CrossRef]
15. Jiang, X.P.; Jin, J.; Wang, Z.Y.; Chen, X.G. A simple dimensionless model for the confinement velocity of the lateral concentrated smoke exhaust in tunnel fire. *Tunn. Undergr. Space Technol.* **2022**, *122*, 104368. [CrossRef]
16. Zhong, W.; Sun, C.P.; Bian, H.T.; Gao, Z.H.; Zhao, J. The plug-holing of lateral mechanical exhaust in subway station: Phenomena, analysis, and numerical verification. *Tunn. Undergr. Space Technol.* **2021**, *112*, 103914. [CrossRef]
17. Zhu, Y.T.; Tang, F.; Huang, Y.J. Experimental study on the smoke plug-holing phenomenon and criteria in a tunnel under the lateral smoke extraction. *Tunn. Undergr. Space Technol.* **2022**, *121*, 104330. [CrossRef]
18. Liu, Q.L.; Xu, Z.S.; Fan, C.G.; Tao, H.W.; Zhao, J.M.; He, L. Experimental and numerical study of plug-holing with lateral smoke exhaust in tunnel fires. *Fire Technol.* **2022**, *1*, 1–21. [CrossRef]
19. Takeuchi, S.; Aoki, T.; Tanaka, F.; Moinuddin, K.A. Modeling for predicting the temperature distribution of smoke during a fire in an underground road tunnel with vertical shafts. *Fire Saf. J.* **2017**, *91*, 312–319. [CrossRef]
20. Ingason, H.; Li, Y.Z.; Lönnermark, A. *Tunnel Fire Dynamics*; Springer: Berlin, Germany, 2014.
21. Hinkley, P.L. The flow of hot gases along an enclosed shopping mall a tentative theory. *Fire Saf. Sci.* **1970**, *807*, 1–17.

22. Ji, J.; Zhong, W.; Li, K.Y.; Shen, X.B.; Zhang, Y.; Huo, R. A simplified calculation method on maximum smoke temperature under the ceiling in subway station fires. *Tunn. Undergr. Space Technol.* **2011**, *26*, 490–496. [CrossRef]
23. Cong, H.Y.; Wang, X.S.; Zhu, P.; Jiang, T.H.; Shi, X.J. Experimental study of the influences of board size and position on smoke extraction efficiency by natural ventilation through a board-coupled shaft during tunnel fires. *Appl. Therm. Eng.* **2018**, *128*, 614–624. [CrossRef]
24. Hong, L.J.; Liu, C.J. Review of research on tunnel fire safety. *Chin. J. Underg. Space Eng.* **2005**, *1*, 149–154.
25. Ji, J.; Tan, T.T.; Gao, Z.H.; Wan, H.X.; Zhu, J.P.; Ding, L. Numerical investigation on the influence of length–width ratio of fire source on the smoke movement and temperature distribution in tunnel fires. *Fire Technol.* **2019**, *55*, 963–979. [CrossRef]
26. Ji, J.; Gao, Z.H.; Fan, C.G.; Sun, J.H. Large Eddy Simulation of stack effect on natural smoke exhausting effect in urban road tunnel fires. *Int. J. Heat Mass Transf.* **2013**, *66*, 531–542. [CrossRef]
27. Ji, J.; Fan, C.G.; Zhong, W.; Shen, X.B.; Sun, J.H. Experimental investigation on influence of different transverse fire locations on maximum smoke temperature under the tunnel ceiling. *Int. J. Heat Mass Transf.* **2012**, *55*, 4817–4826. [CrossRef]

fire

Technical Note

Feasibility Analysis of Cross Passage Ventilation and Smoke Control in Extra-Long Submarine Tunnel

Wenjiao You [1] and Jie Kong [2,*]

[1] Department of Public Security, Zhejiang Police College, Hangzhou 310053, China; youwenjiao@zjjcxy.cn
[2] College of Quality and Safety Engineering, China Jiliang University, Hangzhou 310018, China
* Correspondence: kongjiehao@cjlu.edu.cn; Tel.: +86-15009907342

Abstract: Longitudinal ventilation fans in extra-long submarine tunnels are usually arranged at both ends of the tunnel limited by the tunnel cross section, which is usually hindered by insufficient power caused by extra-long ventilation distances. In this paper, the conception of a ventilation mode is proposed that services the tunnel and cross passages, to provide auxiliary air supply to the main tunnel. Two critical factors have been analyzed on the premise of evacuation safety, which combine to affect the ventilation efficiency in the case of an accident inside the tunnel, these are: air volume within the service tunnel, and cross passage open numbers. FDS simulation software is used to simulate the tunnel model; consider the number of cross passages from one to four; and simulate the service tunnel airflow velocity of 0.7 m/s, 0.75 m/s, 0.85 m/s, 1.0 m/s and 1.3 m/s. The results show that when 1.3 m/s wind speed is applied at both ends of the service tunnel, and three cross passages are operated, 20 MW of fire smoke within the accident tunnel can be effectively controlled; additionally, the wind speed in the cross passage will not hinder the evacuation. The simulation results are verified by ventilation network calculation.

Keywords: extra-long submarine tunnel; smoke control; cross passage; ventilation network

Citation: You, W.; Kong, J. Feasibility Analysis of Cross Passage Ventilation and Smoke Control in Extra-Long Submarine Tunnel. *Fire* **2022**, *5*, 102. https://doi.org/10.3390/fire5040102

Academic Editors: Chuangang Fan and Dahai Qi

Received: 12 June 2022
Accepted: 15 July 2022
Published: 18 July 2022

1. Introduction

In recent years, the extra-long railway tunnel has made travel easier, but also introduces new challenges with regards to safety management [1,2]. When fire breaks out inside a tunnel, poison gas and heat accumulate within the long and narrow space in a short period of time, posing a threat to life [3,4]. Over past decades, tunnel ventilation [5–7] and smoke control [8,9] has been investigated by many researchers, which has proved effective for eliminating back-layering and protecting personnel evacuation. Providing a certain flow velocity from one side of the fire source [10,11], or exhausting smoke through the shaft and flue [12,13], is expected to weaken the fire risk; however, there are still many challenges with long-distance ventilation, including smoke exhaust, evacuation, and emergency rescue. For long-distance ventilation, airflow produced by fans will encounter great resistance within the tunnel and it is hard to form a stable air flow, which may easily cause a disaster, due to the obstructed ventilation.

Tunnel ventilation systems, which provide fresh air to tunnels [14–16], are classified as transverse, semi-transverse and longitudinal. Of these, a longitudinal ventilation system equipped with jet fans is the most widely used, owing to lower costs and easier implementation. Using this ventilation mode, wind flow runs for dozens of kilometers against the wall friction; therefore, a large number of fans are required to provide enough air flow to withstand the huge ventilation loss. Unfortunately, due to the limitation of space within the tunnel, jet fans can only be arranged at both ends of the tunnel, which easily forms a local ultra-high wind pressure, threatening both human and traffic safety. In addition, such an arrangement will also present further train and personnel traffic safety issues, such as excessive ventilation pressure at the end of the tunnel.

Feng et al. [17] studied the smoke control of a subway tunnel fire, and its influencing factors, by use of CFD numerical simulation and full-scale cold-smoke experiment. It was found that the ventilation velocity through the cross passage increased with the shorter distance between train and cross passage. As the fire location was closer to the cross passage, a lower critical velocity was needed for smoke control. Hou et al. [18] carried out research on the critical velocity in a tunnel cross passage through theoretical analysis, full-scale experiment, and numerical simulation. Their results show that when the train blockage is considered in the fire tunnel, the ventilation speed in the cross passage exceeds critical velocity. In the research of Li et al. [19], critical velocity in the cross passage was related to the fireproof door height, fire load and ventilation velocity. They also proposed the dimensionless prediction model of critical velocity in the cross passage, based on their experimental results. Guo et al. [20] used the ventilation network calculation to study wind pressure and measure the tunnel section velocity, exploring tunnel ventilation energy saving technology. Optimal energy saving was proven to reach 43%, which could be applied to practical projects.

Overall, previous studies mainly focused on the critical velocity of fire smoke control in accident tunnels or cross passages, but few studies focused on the design parameters of ventilation systems using cross passages for smoke control, which is worth investigating.

In this paper, a new combination ventilation mode is designed for extra-long tunnels, which adopts longitudinal ventilation of the passenger tunnel, together with auxiliary ventilation in cross passage driven by pressurization of the service tunnel. The twin-tube complementary ventilation system is commonly used in tunnels, because of the advantage of significant reductions in total air volume, and total ventilation power consumption of the two tunnels [21–23]. We will attempt, therefore, to explore the cross passage, which is used as an important part of the ventilation and exhaust system, providing airflow in the case of an accident within extra-long tunnels.

2. Theory and Calculation

2.1. Longitudinal Ventilation Critical Velocity

If a fire occurred in a tunnel under longitudinal ventilation, the critical velocity is defined as the minimum longitudinal airflow velocity, which can limit smoke flow to one side of the fire source. Li et al. [8] carried out a model experiment of tunnel fire to investigate the movement and control of smoke, based on previous research [24]. A set of modified formulas for critical longitudinal velocity can be expressed as:

$$V_c = V_c^* \sqrt{gH} \tag{1}$$

$$V_c^* = \begin{cases} 0.81Q^{*1/3} & , \quad Q^* \le 0.15 \\ 0.43 & , \quad Q^* > 0.15 \end{cases} \tag{2}$$

$$Q^* = \frac{Q}{\rho_0 c_p T_0 g^{1/2} H^{5/2}} \tag{3}$$

where V_c is critical velocity, (m/s); V_c^* is dimensionless critical velocity; Q is heat release rate, (kW); Q^* is dimensionless heat release rate; g is gravitational acceleration, (m/s^2); H is tunnel height, (m); ρ_0 is ambient density, (kg/m^3); c_p is thermal capacity of air, (kJ/kg K); T_0 is ambient temperature, (K).

2.2. Ventilation Parameters

Several factors should be fully considered in ventilation and smoke control in tunnels, such as the pressure of ventilation fans, on-way resistance, local resistance, and fire wind pressure. Among these, fire wind pressure p_f [25], on-way resistance p_λ and local resistance

p_ξ are regarded as ventilation resistance in the design and calculation of the ventilation system, which can be expressed, respectively, as:

$$p_\lambda = \sum_{i=1}^{n} \lambda_i \frac{L_i}{D_i} \frac{\rho}{2} v_i^2 \quad (4)$$

$$p_\xi = \sum_{i=1}^{n} \xi_i \frac{\rho}{2} v_i^2 \quad (5)$$

The ventilation system in the tunnel satisfies the conservation of energy:

$$p_v + p_f + p_\lambda + p_\xi = \frac{1}{2} \rho v_i^2 \quad (6)$$

In order to meet the requirements of smoke control, the average wind velocity v_f at the fire location should exceed the critical longitudinal velocity v_c [8]:

$$v_f \geq v_c \quad (7)$$

where p_λ is on-way resistance, (Pa); p_ξ is local resistance, (Pa); p_v is ventilation pressure, (Pa); p_f is fire wind pressure, (Pa); λ is frictional resistant coefficient; ξ is local resistance coefficient; v_i is airflow velocity, (m/s); v_f is airflow velocity of fire source location, (m/s).

3. Numerical Simulation

3.1. Model Design

In this paper, FDS (Version 6.5.3) codes are used for simulating smoke movement induced by fire [26], which is developed by NIST. The fire tunnel model is based on the Qiongzhou Strait shield tunnel in southern China, which consists of two main tunnels, one service tunnel and multiple cross passages, with a total length of 21 km. To simplify the model, three modules of the tunnel are considered: one main tunnel with a horseshoe section area of 55 m^2 and height of 8.5 m; one service tunnel with a section area of 42.8 m^2; and the cross passages. The total length of the tunnel model is 600 m. The ventilation section covers the whole section of the tunnel, as shown in Figure 1. The section shape is the ventilation space section. A series of temperature and velocity measuring points are arranged at a height of 2 m inside the main tunnel, service tunnel and cross passages, with 1 m interval longitudinally; the velocity and temperature slices are set along the tunnel plotted in Figure 1b. For a high-speed railway tunnel, the heat release rate of 20 MW for train fires is recommended by the 'Code for Design on Rescue Engineering for Disaster Prevention and Evacuation of Railway Tunnel' (TB 10020-2017) [27], which is selected for modeling.

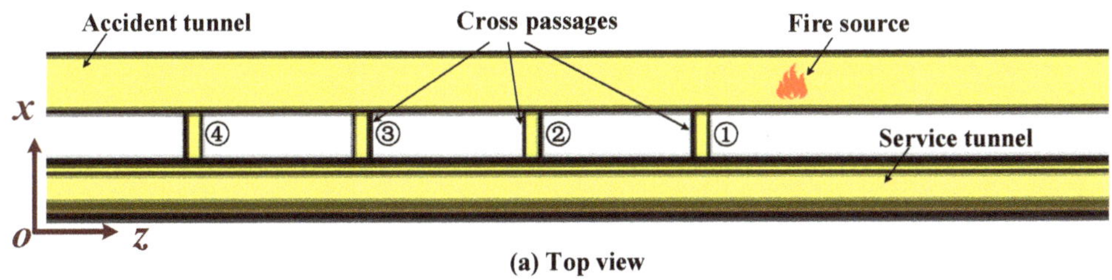

Figure 1. *Cont.*

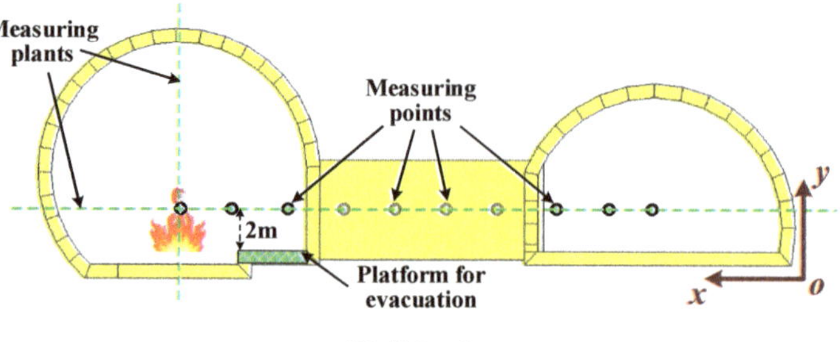

Figure 1. Simulation model: (**a**) Top view; (**b**) Side view.

3.2. Simulated Conditions

Two groups of simulation scenarios are designed, as summarized in Table 1. The first group establishes the model by changing the number of cross passages (CP) from one to four. The interval of cross passages is 50 m, with the cross-sectional area of 12 m^2 and a width × height of 4 m × 3 m. The velocity of longitudinal ventilation generated in an accident tunnel is 2 m/s; the airflow velocity at both ends of the service tunnel is 1.3 m/s. The other group has five conditions, with different ventilation velocities in the service tunnel, set as: 0.70, 0.75, 0.85, 1.00 and 1.30 m/s. Three cross passages remain open. Other parameter conditions are consistent with the first group.

Table 1. Simulated conditions.

Cases No.	Fire Source Location (m)	CP.1 Location (m)	CP.2 Location (m)	CP.3 Location (m)	CP.4 Location (m)	Velocity of Accident Tunnel (m/s)	Velocity of Service Tunnel (m/s)
a01	310	300	–	–	–	2	1.30
a02	310	300	250	–	–	2	1.30
a03	310	300	250	200	–	2	1.30
a04	310	300	250	200	150	2	1.30
b01	310	300	250	200	–	2	0.70
b02	310	300	250	200	–	2	0.75
b03	310	300	250	200	–	2	0.85
b04	310	300	250	200	–	2	1.00
b05/a03	310	300	250	200	–	2	1.30

The fire source is located at 310 m inside the accident tunnel, with a heat release rate of 20 MW, set to reach the maximum value within 1 s, then maintained during the simulation for up to 500 s. Air flow in the service tunnel enters the accident tunnel through cross passages, to raise the longitudinal ventilation velocity to critical velocity. As smoke movement would reach a quasi-steady state after 300 s, the average values from 450 s to 500 s is considered as quasi-steady state data for investigation.

3.3. Grid Sensitivity Analysis

As the gird size is crucial for the reliability of numerical simulation, we chose the value range $D^*/16$ to $D^*/4$, which has been widely used for assessing grid resolution for sensitivity analysis. Here, D^* can be calculated by [26]:

$$D^* = \left(\frac{Q}{\rho_\infty c_p T_\infty \sqrt{g}} \right)^{2/5} \tag{8}$$

A grid size of between 0.2 m and 0.8 m is considered suitable for simulation. In order to obtain a reasonable and accurate mesh size, we took five grid sizes of 0.40 m, 0.50 m, 0.60 m, 0.70 m and 0.80 m, for the numerical simulation stability analysis. The longitudinal ventilation velocity in the tunnel is 3.0 m/s and the fire load is 20 MW. As shown in Figure 2, when the mesh size is less than 0.5 m, the simulation results obtained stability and repeatability, while the CPU time of 0.4 m took 23 h more than 0.5 m (computer configuration: i7-7700, 16 GB RAM). In order to save storage space and computing time, we employed 0.5 m as the mesh size of the numerical simulation model.

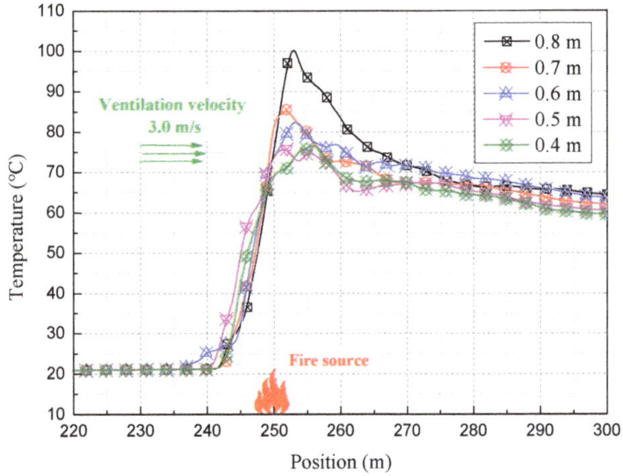

Figure 2. Temperature distribution with different grid sizes.

4. Result and Discussion

4.1. Optimum Number of Cross Passages Open

When a fire occurs in the main tunnel, 2 m/s longitudinal ventilation will be provided, and 1.3 m/s pressurized wind speed will be applied at both ends of the service tunnel. With the number of cross passages open increasing from one to four, the variation of airflow is detected in Figure 3. When only one cross passage is open, the upstream ventilation velocity in the accident tunnel is stable at about 2 m/s, while airflow speed at the intersection of the cross passage and accident tunnel reaches 12 m/s. In addition, the longitudinal ventilation velocity of the main tunnel downstream maintains at 4.7 m/s. As shown in Figure 3a,b–d, the longitudinal wind speed from the tunnel entrance to the first cross passage is basically stable, as the number of cross passages increases. However, velocity is unstable at the intersection of channels, due to chaotic turbulence, gradually stabilizing at 4.5~4.7 m/s when reaching downstream.

In this paper, cross passages are used for the evacuation of people, as well as ventilation. In order to prevent fire smoke from spreading into cross passages, the wind speed in passages has been stipulated to be between 1.5 m/s to 8 m/s by TB 10068-2010 [28]. In fact, as airflow velocity in the cross passage exceeds 5 m/s, evacuation efficiency will be negatively affected; therefore, it is necessary to optimize the number of cross passages open when addressing the issue of smoke control. After simulation, when one cross passage is open, velocity exceeded 8 m/s. When two cross passages are open, the velocity was between 4 m/s and 7 m/s. Average wind speed reached from 1.5 m/s to 5 m/s when CP1, CP2 and CP3 were open. With all four cross passages open, average wind speed in CP3 and CP4 are less than 1.5 m/s, which does not satisfy the safety evacuation requirement. Therefore, when a fire occurs in a tunnel, the three cross passages closest to upstream of the fire source should be open for smoke control.

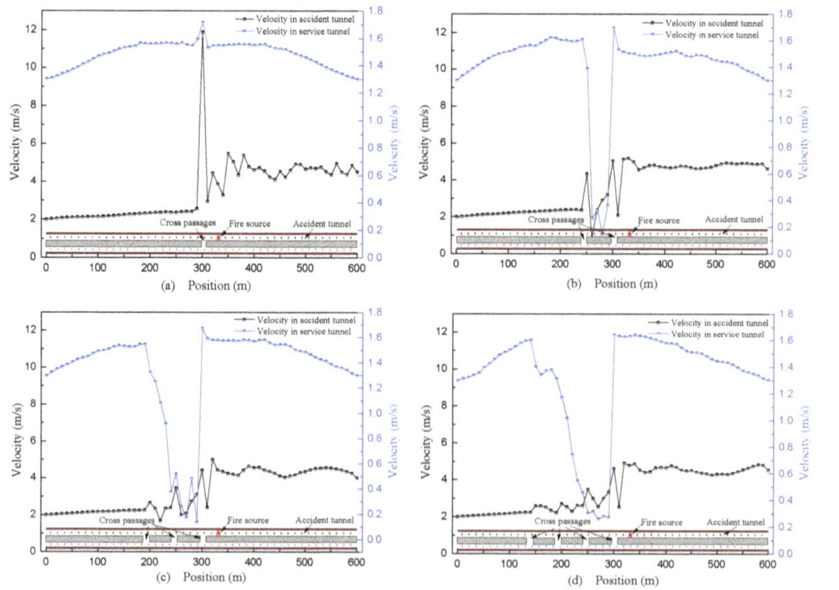

Figure 3. Velocity distribution with different numbers of cross passages open: (**a**) One cross passage open; (**b**) Two cross passages open; (**c**) Three cross passages open; (**d**) Four cross passages open.

4.2. Optimum Ventilation Quantity of Service Channel

The distribution of smoke movement and airflow velocity, with three cross passages open, were analyzed at ventilation velocities of 0.7 m/s, 0.75 m/s, 0.85 m/s, 1.0 m/s and 1.3 m/s, at both ends of the service tunnel. Figure 4 compares the accident tunnel velocities at a height of 2 m, under a service tunnel wind speed from 0.7 m/s to 1.3 m/s. As reflected in the figures, velocities near the fire source increased with service tunnel wind speed; velocities in the cross passages were all less than 8 m/s, under different working conditions. Of note, the average wind speed in some cross passages was less than 1.5 m/s in airflow velocity conditions of 0.7 m/s, 0.75 m/s and 0.85 m/s; therefore, it is necessary to reconsider tunnel ventilation design systems, in order to ensure the effectiveness of fire smoke control in tunnel ventilation.

4.3. Ventilation Network Verification

In tunnel ventilation systems, network calculation can capture overall mean flow parameters, such as velocity and pressure, with a one-dimensional flow regime [29]. Two typical concepts, node and branch, will be discussed. Both tunnels and cross passages are characterized as branches, while the intersection point of branches are described as nodes. Figure 5 shows typical tunnels with a three cross passage ventilation network, which includes a main loop, ten nodes and fourteen branches.

The ventilation airflow at the *m*th node in tunnels follows the mass balance Equation (9), and the pressure balance equation for the *n*th branch can be represented as Equation (10). The on-way resistance and local resistance will meet Equations (4) and (5), respectively. In addition, fire exists as resistance in branch and its wind pressure is set at 5 Pa [25].

$$\sum M_m = 0 \tag{9}$$

$$\sum p_n = 0 \tag{10}$$

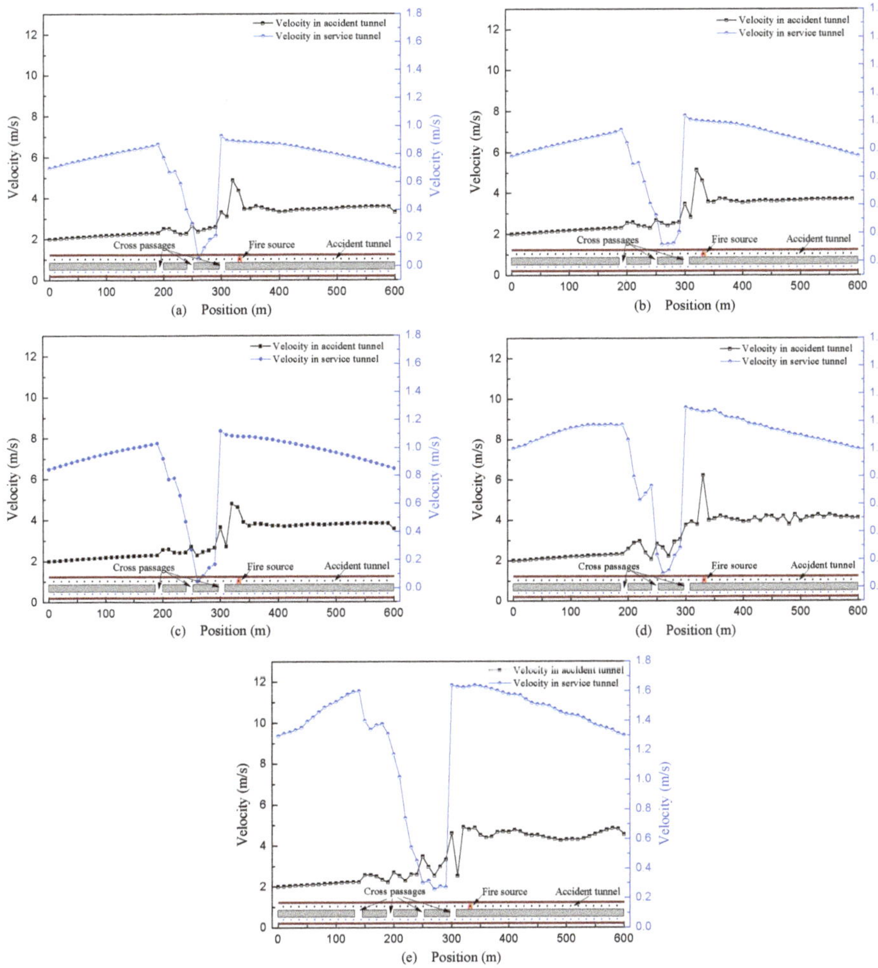

Figure 4. Velocities in both accident tunnel and service tunnel: The velocity at both ends of service tunnel is (**a**) 0.70 m/s; (**b**) 0.75 m/s; (**c**) 0.80 m/s; (**d**) 1.00 m/s; (**e**) 1.30 m/s.

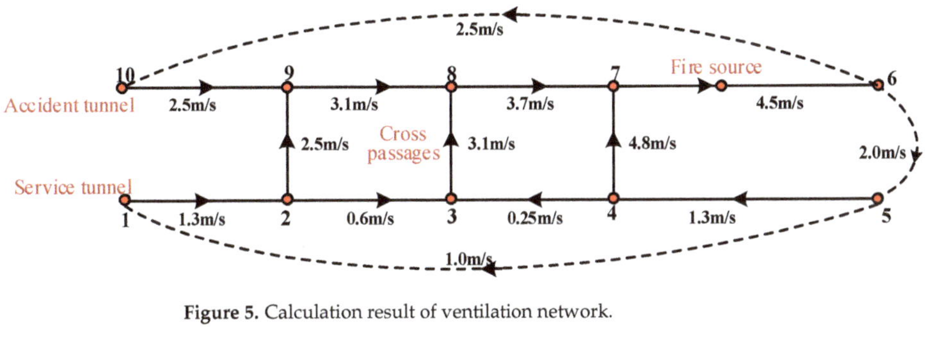

Figure 5. Calculation result of ventilation network.

Figure 5 represents the calculation results and data of the ventilation network. At both ends of the accident tunnel, ventilation pressure of 210 Pa is applied in the same direction, and ventilation pressure of 135 Pa with opposite direction is set in the service tunnel. The ventilation volume of the CP.1, CP.2 and CP.3 are 29.9 m^3, 36.9 m^3 and 45.5 m^3, respectively; the average airflow velocity are 2.5 m/s, 3.1 m/s and 4.8 m/s. The fire source is located in the 10th branch, with ventilation velocity of 4.5 m/s, exceeding the longitudinal critical velocity of 3.5 m/s. The calculation results of the ventilation network agree with the numerical simulation. Therefore, it is feasible to use the service tunnel to pressurize ventilation through three open cross passages, in order to limit smoke movement in the case of a 20 MW fire in the main tunnel.

5. Conclusions

In this paper, a ventilation mode with service tunnel and cross passages for auxiliary air supply and smoke control was studied using FDS simulation. The study proposed a ventilation and smoke exhaust scheme, and solved the problem of longitudinal ventilation in extra-long submarine tunnels. Our main conclusions are as follows:

(1) In the case of a 20 MW fire, the longitudinal ventilation velocity of 2 m/s will be applied in the accident tunnel, and the airflow speed of 1.3 m/s will be supplied at both ends of the service tunnel, together with three cross passages open to provide airflow, which can effectively control the fire smoke and improve the human evacuation.

(2) A ventilation network model is established according to the design parameters of extra-long tunnels. The calculation results show that the longitudinal wind speed at the fire source reaches 4.5 m/s, exceeding the critical velocity for smoke control of 3.5 m/s, which is in accordance with the numerical simulation results. In theory, cross passage pressurized air supply technology is proved feasible.

It should be noted that the distance between cross passages, and the angle between cross passages with main tunnel, will influence ventilation efficiency. Therefore, further experiments and simulations are needed, and parameters should be extended in order to investigate the optimum ventilation scheme. In this paper, design parameters are closely related to the tunnel structure, which are not necessarily applicable to other projects. With regards to ventilation design, the methods and ideas highlighted in this paper are significant; perhaps other tunnels need similar structural models and calculation conditions.

Author Contributions: Conceptualization, W.Y. and J.K.; methodology, W.Y.; software, J.K.; validation, W.Y. and J.K.; formal analysis, W.Y.; investigation, W.Y.; resources, W.Y.; data curation, W.Y.; writing—original draft preparation, J.K.; writing—review and editing, J.K.; visualization, W.Y.; supervision, W.Y.; project administration, J.K.; funding acquisition, J.K. All authors have read and agreed to the published version of the manuscript.

Funding: This research was funded by Philosophy and Social Science Planning of Zhejiang, grant number 20NDJC199YB; National Natural Science Foundation of Zhejiang, grant number GF22F030254.

Institutional Review Board Statement: Not applicable.

Informed Consent Statement: Not applicable.

Data Availability Statement: Not applicable.

Conflicts of Interest: The authors declare no conflict of interest.

References

1. Guo, X.X.; Pan, X.H.; Wang, Z.L.; Yang, J.; Hua, M.; Jiang, J.C. Numerical simulation of fire smoke in extra-long river-crossing subway tunnels. *Tunn. Undergr. Space Technol.* **2018**, *82*, 82–98. [CrossRef]
2. Xu, Z.S.; You, W.J.; Kong, J.; Cao, H.H.; Zhou, C. A study of fire smoke spreading and control in emergency rescue stations of extra-long railway tunnels. *J. Loss Prev. Process Ind.* **2017**, *49*, 155–161. [CrossRef]
3. Capote, J.A.; Alvear, D.; Abreu, O.; Cuesta, A. Analysis of evacuation procedures in high speed trains fires. *Fire Saf. J.* **2012**, *49*, 35–46. [CrossRef]
4. Kong, J.; Xu, Z.S.; You, W.J. Research on Evacuation of Train Fire in Extra-Long Channel Tunnel. In Proceedings of the Asia-Oceania Symposium on Fire Science and Technology, Taipei, China, 22 October 2018. [CrossRef]
5. Wang, F.; Wang, M.N.; He, S.; Zhang, J.S.; Deng, Y.Y. Computational study of effects of jet fans on the ventilation of a highway curved tunnel. *Tunn. Undergr. Space Technol.* **2010**, *25*, 382–390. [CrossRef]

6. Wang, F.; Wang, M.N.; Wang, Q.Y. Numerical study of effects of deflected angles of jet fans on the normal ventilation in a curved tunnel. *Tunn. Undergr. Space Technol.* **2012**, *31*, 80–85. [CrossRef]
7. Li, Y.Z.; Chen, T.; Xu, Z.S.; Kong, J.; Wang, M.Q.; Fan, C.G. Influence of winding wall on the entrainment characteristics of air jet in curved road tunnels. *Tunn. Undergr. Space Technol.* **2019**, *90*, 330–339. [CrossRef]
8. Li, Y.Z.; Lei, B.; Ingason, H. Study of critical velocity and backlayering length in longitudinally ventilated tunnel fires. *Fire Saf. J.* **2010**, *45*, 361–370. [CrossRef]
9. Wang, F.; Wang, M.N.; Carvel, R.; Wang, Y. Numerical study on fire smoke movement and control in curved road tunnels. *Tunn. Undergr. Space Technol.* **2017**, *67*, 1–7. [CrossRef]
10. Ingason, H.; Li, Y.Z. Model scale tunnel fire tests with longitudinal ventilation. *Fire Saf. J.* **2010**, *45*, 371–384. [CrossRef]
11. Tang, F.; Cao, Z.L.; Palacios, A.; Wang, Q. A study on the maximum temperature of ceiling jet induced by rectangular-source fires in a tunnel using ceiling smoke extraction. *Int. J. Therm. Sci.* **2018**, *127*, 329–334. [CrossRef]
12. Tang, F.; He, Q.; Mei, F.Z.; Wang, Q.; Zhang, H. Effect of ceiling centralized mechanical smoke exhaust on the critical velocity that inhibits the reverse flow of thermal plume in a longitudinal ventilated tunnel. *Tunn. Undergr. Space Technol.* **2018**, *82*, 191–198. [CrossRef]
13. Wan, H.X.; Gao, Z.H.; Han, J.Y.; Ji, J.; Ye, M.J.; Zhang, Y.M. A numerical study on smoke back-layering length and inlet air velocity of fires in an inclined tunnel under natural ventilation with a vertical shaft. *Int. J. Therm. Sci.* **2019**, *138*, 293–303. [CrossRef]
14. Chen, T.Y.; Lee, Y.T.; Hsu, C.C. Investigations of piston-effect and jet fan-effect in model vehicle tunnels. *J. Wind Eng. Ind. Aerodyn.* **1998**, *73*, 99–110. [CrossRef]
15. Bari, S.; Naser, J. Simulation of smoke from a burning vehicle and pollution levels caused by traffic jam in a road tunnel. *Tunn. Undergr. Space Technol.* **2005**, *20*, 281–290. [CrossRef]
16. Colella, F.; Rein, G.; Carvel, R.; Reszka, P.; Torero, J.L. Analysis of the ventilation systems in the Dartford tunnels using a multi-scale modelling approach. *Tunn. Undergr. Space Technol.* **2010**, *25*, 423–432. [CrossRef]
17. Feng, S.; Li, Y.; Hou, Y.; Li, J.; Huang, Y. Study on the critical velocity for smoke control in a subway tunnel cross-passage. *Tunn. Undergr. Space Technol.* **2020**, *97*, 103234. [CrossRef]
18. Hou, Y.S.; Li, Y.F.; Li, J.M. Study on Train Obstruction Effect on Smoke Control near Tunnel Cross-Passage. In Proceedings of the Eighth International Seminar on Fire and Explosion Hazards (ISFEH8), Hefei, China, 25–28 April 2016; pp. 85–93. [CrossRef]
19. Li, Y.Z.; Lei, B.; Ingason, H. Theoretical and Experimental Study of Critical Velocity for Smoke Control in a Tunnel Cross-Passage. *Fire Technol.* **2013**, *49*, 435–449. [CrossRef]
20. Guo, C.; Xu, J.; Yang, L.; Guo, X.; Zhang, Y.; Wang, M. Energy-Saving Network Ventilation Technology of Extra-Long Tunnel in Climate Separation Zone. *Appl. Sci.* **2017**, *7*, 454. [CrossRef]
21. Zhang, H.; Sun, J.C.; Lin, F.; Wang, L. Optimization on energy saving ventilation of gallery-type combined construction shaft exhaust in extra long tunnel. *Procedia Eng.* **2017**, *205*, 1777–1784. [CrossRef]
22. Ren, R.; Xu, S.S.; Ren, Z.D.; Zhang, S.Z.; Wang, H.; Wang, X.L.; He, S.Y. Numerical investigation of particle concentration distribution characteristics in twin-tunnel complementary ventilation system. *Math. Probl. Eng.* **2018**, *2018*, 1329187. [CrossRef]
23. Wang, Y.Q.; Xu, S.S.; Ren, R.; Zhang, S.Z.; Ren, Z.D. Application of the twin-tube complementary ventilation system in large-slopping road tunnels in China. *Int. J. Vent.* **2020**, *19*, 63–82. [CrossRef]
24. Oka, Y.; Atkinson, G.T. Control of smoke flow in tunnel fires. *Fire Saf. J.* **1995**, *25*, 305–322. [CrossRef]
25. Zhang, T.B.; Zhang, Z.D. Discussion about the computation method for the fire-caused wind pressure during a tunnel fire. *Mod. Tunn. Technol.* **2010**, *47*, 17–21. (In Chinese)
26. McGrattan, K.B.; Hostikka, S.; McDermott, R.; Floyd, J.E. *Fire Dynamics Simulator (Version 6): Technical Reference Guide*; National Institute of Standards and Technology: Gaithersburg, MD, USA, 2014.
27. *TB 10020-2017*; Code for Design on Rescue Engineering for Disaster Prevention and Evacuation of Railway Tunnel. China Railway Publishing House: Beijing, China, 2017. (In Chinese)
28. *TB 10068-2010*; Code for Design on Operating Ventilation of Railway Tunnel. China Railway Publishing House: Beijing, China, 2010. (In Chinese)
29. Du, T.; Yang, D.; Peng, S.N.; Xiao, Y.M.; Zhang, F. Longitudinal ventilation for smoke control of urban traffic link tunnel: Hybrid field-network simulation. *Procedia Eng.* **2014**, *84*, 586–594. [CrossRef]

fire

Technical Note

Study on the Effect of Blockage Ratio on Maximum Smoke Temperature Rise in the Underground Interconnected Tunnel

Zhisheng Xu [1,2], Yaxing Zhen [1,2], Baochao Xie [1,2], Sylvain Marcial Sakepa Tagne [1], Jiaming Zhao [1,2] and Houlin Ying [1,2,*]

1 School of Civil Engineering, Central South University (CSU), Changsha 410075, China
2 National Engineering Research Center of High-Speed Railway Construction Technology, Central South University, Changsha 410075, China
* Correspondence: evelynlin@csu.edu.cn; Tel.: +86-731-8265-5177

Abstract: The model-scale tunnel is used in this investigation to analyze the maximum smoke temperature rise of the interconnected tunnel for various longitudinal ventilation velocities, blockage ratios, and heat release rates where the fire is at the confluence of the underground interconnected tunnel. The results showed that the longitudinal ventilation velocities of both the ramp upstream of the fire source and the adjacent ramp influenced the maximum temperature rise under the underground interconnected tunnel, and the ventilation of both ramps jointly affected the maximum temperature rise. The change in the maximum temperature rise depends on who is more affected by the longitudinal ventilation velocity or the vehicle blockage ratio. As the longitudinal ventilation velocity in the interconnected tunnel increases, the convective heat transfer near the fire source increases, resulting in a decrease in the maximum temperature rise, and the effect of the blockage ratio on the maximum temperature rise is reduced. In this paper, a maximum temperature rise prediction model suitable for the case of blockage in the interconnected tunnel is proposed.

Keywords: underground interconnected tunnel; maximum temperature rise; blockage ratio; longitudinal ventilation velocity

Citation: Xu, Z.; Zhen, Y.; Xie, B.; Sakepa Tagne, S.M.; Zhao, J.; Ying, H. Study on the Effect of Blockage Ratio on Maximum Smoke Temperature Rise in the Underground Interconnected Tunnel. *Fire* **2023**, *6*, 50. https://doi.org/10.3390/fire6020050

Academic Editors: Chuangang Fan and Dahai Qi

Received: 17 December 2022
Revised: 12 January 2023
Accepted: 30 January 2023
Published: 31 January 2023

1. Introduction

Urban ground traffic congestion is a common problem faced in the development of cities. Building "three-dimensional transportation" in cities has gradually emerged to relieve traffic pressure in large cities. The underground interconnected infrastructure is used to change the three-dimensional traffic line from the ground to the underground [1]. The structure of an underground interconnected tunnel is complex, and the consequences are serious once a fire occurs [2,3].

Alpert [4] analyzed the maximum temperature rise generated by the structure fire in the absence of wind to establish a forecast model. Heskestad and Delichatsios [5] proposed another classical equation for calculating the maximum temperature rise through the dimensionless heat release rate (HRR). Moreover, researchers investigated the impact of parameters such as ventilation velocity, tunnel slope, and smoke exhausting methods on the maximum temperature rise [6–10].

Highway tunnel ventilation methods primarily include longitudinal, transverse, and semitransverse ventilation [11–13]. In terms of cost, longitudinal ventilation continues to be the most commonly used method of smoke control in the construction of tunnel projects around the world [14]. In 2003, Kurioka [15], a Japanese scholar, conducted experiments in three different scaling models and combined the tunnel section aspect ratio, HRR, and longitudinal ventilation velocity to propose the maximum temperature rise prediction equation:

$$\frac{\Delta T_m}{T_0} = \gamma\left(\frac{\dot{Q}^{*2/3}}{Fr^{1/3}}\right)^{\varepsilon} \quad (1)$$

$$\dot{Q}^{*2/3}/Fr^{1/3} < 1.35, \quad \gamma = 1.77, \varepsilon = 6/5$$
$$\dot{Q}^{*2/3}/Fr^{1/3} \geq 1.35, \quad \gamma = 2.54, \varepsilon = 0 \tag{2}$$

where the dimensionless HRR and the Froude number were written as:

$$\dot{Q}^{*} = \frac{\dot{Q}}{\rho c_p T_0 g^{1/2} H^{5/2}}, Fr = \frac{V^2}{gH} \tag{3}$$

where:

ΔT_m is the maximum temperature rise (K);
T_0 is the ambient temperature (K);
$\dot{Q}$ is the heat release rate (kW);
γ, ε is the experimental constant;
ρ is the air density (kg/m^3);
c_p is the specific heat at constant pressure (kJ/(kg·K));
g is the gravitational acceleration (m/s^2);
H is the tunnel height (m);
V is the longitudinal ventilation velocity (m/s).

Li [16] discovered that Kurioka's formula could not accurately predict the maximum temperature rise at a shallow velocity. Thus, Li improved the formula concerning the maximum temperature rise with longitudinal ventilation and without longitudinal ventilation as follows:

$$\Delta T_{\mathrm{m}} = \begin{cases} \frac{\dot{Q}}{V b_{fo}^{1/3} H_{ef}^{5/3}}, & V' > 0.19 \\ 17.5 \frac{\dot{Q}^{2/3}}{H_{ef}^{5/3}}, & V' \leq 0.19 \end{cases} \tag{4}$$

where:

$\dot{Q}_{es}$ is the total HRR (kW);
b_{fo} is the fire source's radius (m);
H_{ef} is the vertical distance above the bottom of the fire source (m);
V' is the dimensionless longitudinal ventilation velocity.

With the construction of urban underground interconnected tunnels, the tunnels need to connect with the ground road or other underground roads, which makes the underground interconnected tunnel more complex and the traffic flow larger. It is simple to create a blockage once a fire has started.

The maximum tunnel temperature rise in the situation of a single-pipe tunnel blockage has been the subject of research. Li et al. [17] and Luo et al. [18] investigated the relationship between the maximum temperature rise and the blockage ratio during longitudinal ventilation. The maximum temperature is reduced as the blockage ratio rises. Tang [19] investigated the impact of the distance between the obstruction site and the fire source on the maximum temperature rise when the longitudinal ventilation was adjusted using a model-scale test. Wang [20] analyzed the maximum temperature rise at different velocities and heat release rates (HRRs) and provided a formula in the event of a blockage. In Kayili's [21] study, the maximum temperature rise and the blockage ratio were correlated. The results showed that the blockage ratio was proportional to the maximum temperature rise for a particular ventilation velocity. In addition, Li et al. [22] experimentally investigated the influence of blockage on the maximum smoke temperature and revised the Kurioka model by introducing the blockage ratio coefficient ϕ.

$$\frac{\Delta T_{\mathrm{m}}}{T_0} = \gamma \left[\frac{\dot{Q}^{2/3}}{Fr^{1/3}} \cdot (1-\phi)^{5/2} \right]^{\varepsilon} \tag{5}$$

where:

ϕ is the blockage ratio (%).

Due to the existence of the interconnected tunnel, the ventilation airflow velocity at the bifurcation is inconstant, and the smoke diffusion characteristics and control under fire conditions are very different from those of ordinary single-pipe tunnels. Huang et al. [23] investigated the influence of bifurcated angles, HRRs, and velocities on maximum temperature rise. A correlation equation for the maximum temperature variation under the bifurcated tunnel's ceiling was proposed. Whether the fire source was upstream or at the bifurcation point, Chen et al. [24] found that as the velocity increased, the maximum temperature rise in the main tunnel decreased linearly.

Previous research focused on the bifurcation angle, longitudinal ventilation velocity, and fire source location, most of which employed a single-pipe straight tunnel. However, in the interconnected tunnel, it has not yet been possible to develop a predicted correlation of the maximum temperature rise that takes into account the coupling impact of various blockage ratios, ventilation velocities, and HRRs. In this work, a prediction method is proposed along with physical model tests to evaluate the effects of the blockage ratio, ventilation velocity, and HRR on smoke maximum temperature rise.

2. Experimental Setup

2.1. Model-Scale Tunnel

A 1:30 underground interconnected tunnel model was constructed. The interconnected tunnel consisted of multiple curved and straight tunnel sections, a ramp D length of 6 m, and a ramp C length of 21.5 m, and the cross-sectional dimension of the ramp was 0.32 m (W) × 0.22 m (H). The tunnel sidewalls were composed of 6 mm thick fire-proof glass, while the tunnel ceilings were composed of 3 mm thick fire-proof plates. The model-scale tunnel is shown in Figure 1. Ramp C and ramp D traffic flows were at a confluence at the bifurcated point.

Figure 1. Model-scale underground interconnected tunnel.

The underground interconnected tunnel was made up of 8 multiple curved tunnel segments, as shown in Figure 2, with Ramp C and D forming a merge at the splitting point. The tunnel had no slopes. Table 1 shows the length and radius of curvature of each segment.

The thermocouples arrangement is shown in Figure 2. They were placed 0.01 m under the tunnel ceiling, and the longitudinal distance was 0.15 m. For a more accurate measurement of the maximum temperature rise of the tunnel, the thermocouple array was arranged at the confluence. There were a total of 108 horizontal and longitudinal intervals of 0.06 m. Propane gas was used as the fire source fuel in the test. Propane has the characteristics of low pollution and controllable combustion. At present, a large number of scholars have applied it to the test of reduced-size fire. In the experiment, the mass flow rate of fuel was measured and controlled by a mass flowmeter to control the HRR of the fire source. The burner's inner diameter was 0.03 m × 0.03 m. The fire source was situated near the bottom of the tunnel, in the center of the D ramp, and was 0.02 m high.

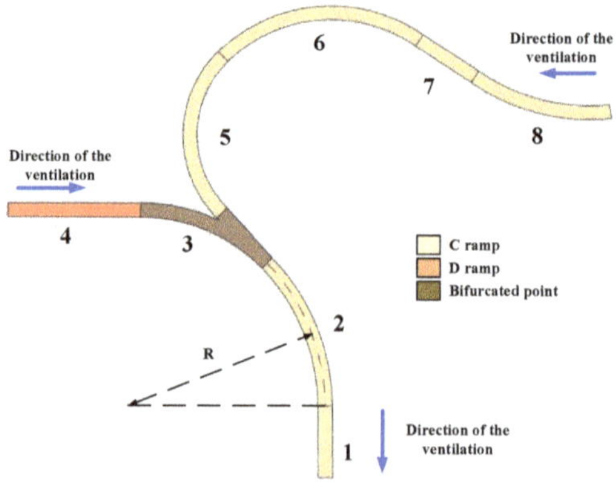

Figure 2. Schematic diagram of the different sections of the underground interconnected tunnel.

Table 1. Parameters of different segments of the underground interconnected tunnel.

Ramp	Number	Curvature Radius R (m)	Length (m)
C	1	∞	3.00
C	2	4.35	3.41
D	3(bifurcated point)	-	-
D	4	∞	3.00
C	5	2.50	3.93
C	6	2.50	4.88
C	7	∞	1.50
C	8	4.50	3.22

This test adopted the longitudinal smoke exhaust mode. The jet fans were 0.185 m long, had an outlet diameter of 37 mm, and an operating voltage of 4.5 V–12 V. As shown in Figure 1, a group of jet fans was arranged at the entrance of ramp C and ramp D, and in each group with two jet fans, wind was blown from the entrance of ramp C and ramp D to the exit of ramp C. The fans were fixed in the tunnel by customized brackets, and the tunnel's ceiling was 0.045 m from the jet fans' center. The lateral distance between the two jet fans was 0.11 m, which achieved the triple outlet diameter of the jet fans. The adjustable direct current (D.C.) power was used to manage the longitudinal ventilation velocities. The locations for measuring the airflow velocity were in the center of the tunnel cross-section, but the blockage measurement points were above the obstruction.

The selection of the HRR was the most critical aspect of the tunnel fire test. The HRRs were set to be 5 MW, 8 MW, 12 MW, 16 MW, and 20 MW, which should be 1.01 kW, 1.83 kW, 2.43 kW, 3.24 kW, and 4.05 kW, respectively, when generalized on a small scale by using the Froude number.

As the blockage vehicle at the upstream fire source, three typical vehicles were chosen. The distance between the fire source and the vehicle was 0.1 m. For the convenience of calculation, the vehicles are simplified as rectangular bodies, and the width and height of each vehicle were 0.06 m (W) × 0.055 m (H), 0.08 m (W) × 0.10 m (H), and 0.08 m (W) × 0.13 m (H). The vehicle blockage length was 0.8 m, and the blockage area was the area of the same type of two vehicles side by side, as shown in Figure 3. Therefore, the tunnel blockage ratios (φ) were 0%, 10%, 20%, and 30%, respectively.

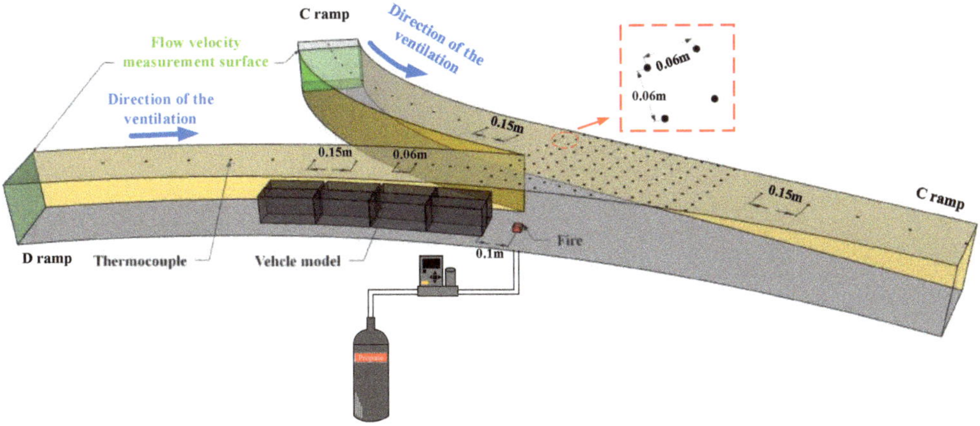

Figure 3. The layout of test points for the model-scale test.

2.2. Experimental Conditions

Table 2 summarizes the parameters of the model-scale tunnel experiment. The blockage ratios (ϕ) were set to be 0%, 10%, 20%, and 30%. In addition, five different HRRs (1.01 kW, 1.83 kW, 2.43 kW, 3.24 kW, and 4.05 kW) and six ventilation methods (velocity was based on the velocity at ϕ = 0%) were considered. A total of 120 test conditions were developed.

Table 2. Experimental conditions.

Test NO.	Velocity Scheme	HRR (kW)	Blockage Ratio (%) ϕ	Ramp C Velocity (m/s) V_C	Ramp D Velocity (m/s) V_D
1–20	Scheme 1	1.01, 1.83, 2.43, 3.24, 4.05	0, 10, 20, 30	0.42	0.38
21–40	Scheme 2			0.41	0.45
41–60	Scheme 3			0.40	0.61
61–80	Scheme 4			0.44	0.55
81–100	Scheme 5			0.46	0.85
101–120	Scheme 6			0.65	0.96

3. Results and Discussion

3.1. Variations in Longitudinal Ventilation Velocities in the Interconnected Tunnel

Figure 4 shows the variations in the longitudinal ventilation velocities with different blockage ratios for the same velocity scheme. When velocity Scheme 1 was used for the interconnected tunnel, the velocity at ramp C remained essentially constant, as the blockage ratio rose. The velocity of ramp D showed a decreasing trend. The velocity dropped from 0.38 m/s to 0.16 m/s and changed significantly. When velocity Scheme 6 was used for the interconnected tunnel, as the blockage ratio rose, the velocity at ramp C varied between 0.6 m/s and 0.8 m/s. The velocity of ramp D showed a decreasing trend. The velocity dropped from 0.96 m/s to 0.78 m/s, which also changed significantly. Even though the interconnected tunnel used the same velocity scheme, the actual velocity changed as a result of the blockage in the tunnel, among which ramp D changed greatly. Therefore, the longitudinal ventilation velocities of ramp C and ramp D in this paper were based on the actual velocity measured by the flow velocity measuring point in the interconnected tunnel, as shown in Table 3.

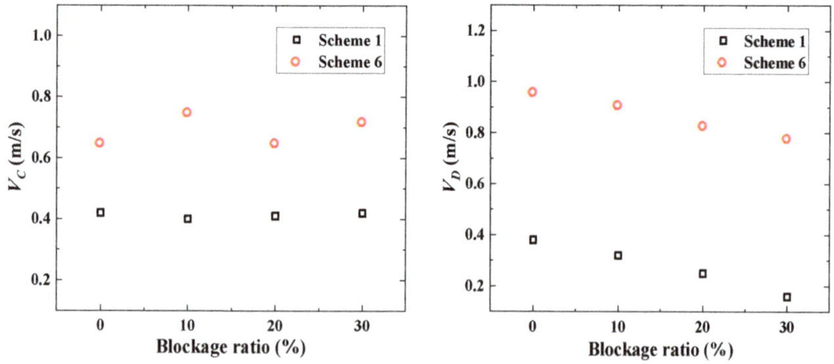

(a) (b)

Figure 4. Variation in longitudinal ventilation velocities with different blockage ratios under the same velocity scheme: (**a**) ramp C velocity; (**b**) ramp D velocity.

Table 3. Longitudinal ventilation velocity measured values of ramp C and ramp D.

Test NO.	Longitudinal Ventilation Velocities (m/s)								
	ϕ	0%		10%		20%		30%	
	Ramp	C	D	C	D	C	D	C	D
1–20		0.42	0.38	0.40	0.32	0.41	0.25	0.42	0.16
21–40		0.41	0.45	0.28	0.47	0.37	0.44	0.30	0.40
41–60		0.40	0.61	0.35	0.65	0.34	0.62	0.33	0.55
61–80		0.44	0.55	0.40	0.60	0.46	0.46	0.47	0.41
81–100		0.46	0.85	0.38	0.87	0.43	0.73	0.43	0.68
101–120		0.65	0.96	0.75	0.91	0.65	0.83	0.72	0.78

Figure 5 shows the sum of the ramps C and D velocities with the blockage ratios. In the interconnected tunnel, both ramps C and D had longitudinal ventilation. When the blockage ratio increased, the velocity total (V_{C+D}) of ramps C and D dropped to some amount. According to Figure 3, this was because V_D decreased as the blockage ratio increased.

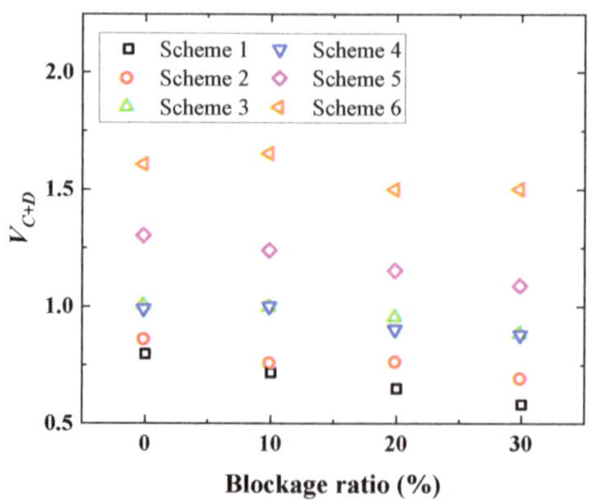

Figure 5. Variation in the sum of longitudinal ventilation velocities with blockage ratio for ramps C and D.

Ramp C and ramp D had longitudinal ventilation in the interconnected tunnel. Figure 6 selects the data for ramp C with similar velocities and ramp D with different velocities. When ϕ = 20%, $\dot{Q}$ = 4.05 kW, ramp C's velocities were comparable, and ramp D's velocities varied from 0.25 m/s to 0.73 m/s, and the maximum temperature rise drastically dropped from 501 K to 267 K. The maximum temperature rise varied dramatically when the velocities at ramp D changed while the blockage ratio was constant and the velocities at ramp C were similar. This may be related to the fire source located on ramp D, which made ramp D velocities have a greater impact on the smoke.

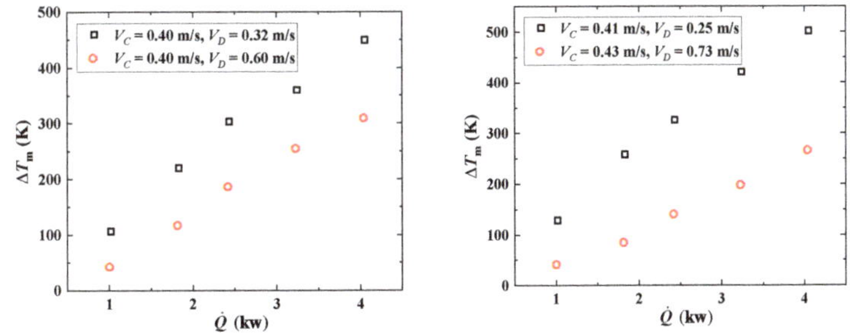

(a) (b)

Figure 6. Variation in maximum temperature rise with HRR under similar velocities on ramp C: (**a**) ϕ = 10%; (**b**) ϕ = 20%.

Figure 7 selects the maximum temperature rise data when the velocities of the D ramp were similar and the velocities of the C ramp were different. When ϕ = 20%, $\dot{Q}$ = 4.05 kW, the ramp C velocities were 0.37 m/s and 0.46 m/s, and the maximum temperature was 405 K while the minimum was 317 K. The longitudinal ventilation velocity in ramp C also caused a change in the maximum temperature rise.

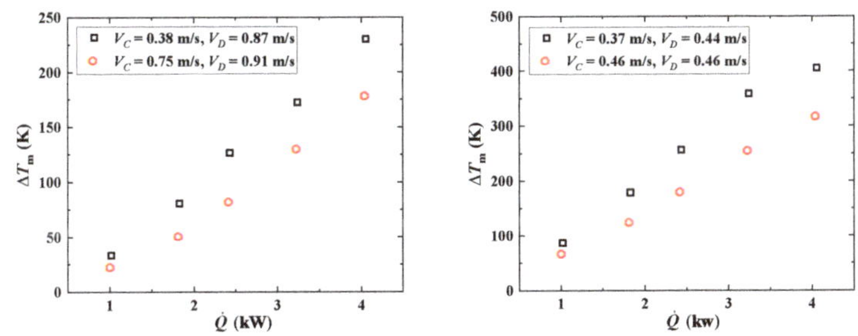

(a) (b)

Figure 7. Variation in maximum temperature rise with HRR under similar velocities on ramp D: (**a**) ϕ = 10%; (**b**) ϕ = 20%.

The velocities of ramp C and ramp D jointly affected the maximum temperature rise. Any change in the velocities of one of the ramps would cause the maximum temperature rise to vary. The influence of ramp D was greater, which could have been because the fire source was positioned on ramp D and influenced it significantly.

From Figures 6–8, with an increasing HRR, the maximum temperature rise increased. When the velocity increased for both ramps C and D, $\dot{Q}$ = 4.05 kW, the maximum temperature rise fell from 405 K to 107 K. The addition of HRR gradually reduced the increase in the maximum temperature rise.

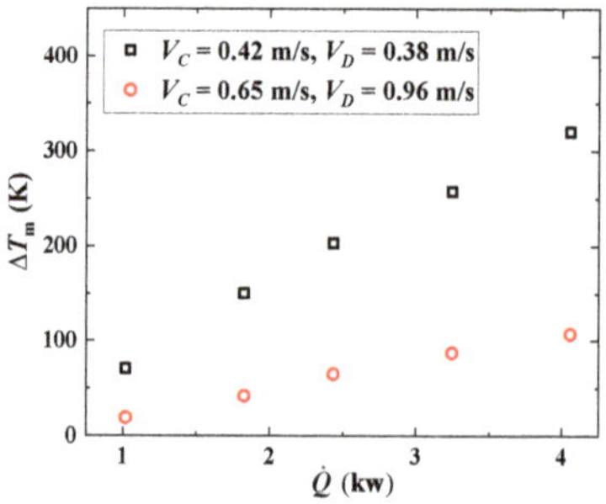

Figure 8. The maximum temperature rise varies with HRR (ϕ = 0%).

3.2. Variation in Maximum Temperature Rises with Blockage Ratios in the Interconnected Tunnel

At a certain ventilation velocity, the maximum temperature rise increased with the blockage ratio for the 0%, 10%, and 20% blockage ratios, as shown in Figure 9. Behind the blockage, the wind created a vortex area. Because the wind flow created a vortex area behind the obstruction when the blockage was located upstream of the fire, it resulted in a slowed local velocity and enhanced thermal feedback of the fire source, as well as an increase in the maximum temperature. The length and size of the vortex would have steadily risen as the blocking ratio increased [21]. When the blockage ratio was 30%, it was rather high; when the blockage ratio was increased, velocity increased, but the maximum temperature rise was reduced. Who was more impacted by the vehicle blockage ratio or velocity determines the maximum temperature rise in the interconnected tunnel.

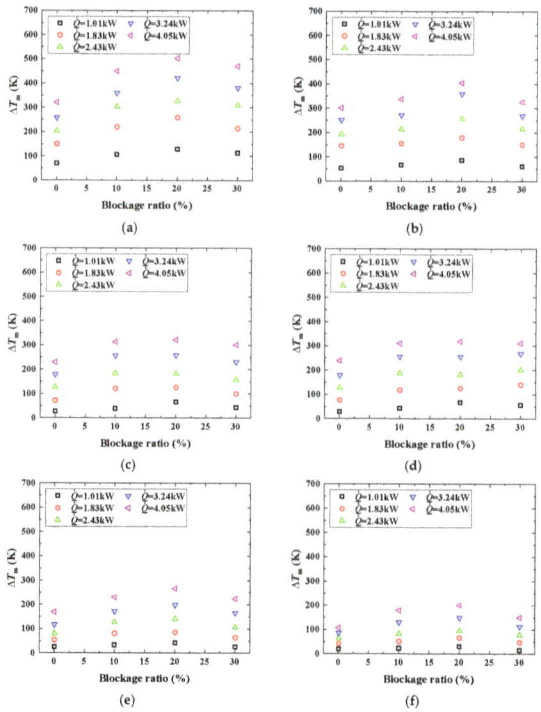

Figure 9. The maximum temperature rise varies with blockage ratio: (**a**) Scheme 1; (**b**) Scheme 2; (**c**) Scheme 3; (**d**) Scheme 4; (**e**) Scheme 5; (**f**) Scheme 6.

According to the analysis of 3.1, the velocities of ramp C and ramp D jointly affected the maximum temperature rise of the interconnected tunnel. The change in the maximum temperature rise from velocity Scheme 1 to 6 shows that as the velocity increased in the interconnected tunnel, convective heat transfer near the fire source increased, resulting in a decrease in temperature and a decrease in the effect of the blockage ratio on the maximum temperature rise.

3.3. *Maximum Temperature Rise Prediction Model for Interconnected Tunnel*

The experimental data were processed by the maximum temperature rise prediction Equation (6) proposed by Li [16]. The sum of the velocities of ramps C and D was used as the characteristic velocity. The maximum temperature rise data fit well, as shown in Figure 10. It demonstrated how the velocities of ramp C and ramp D had an impact on the maximum temperature rise. However, the experimental results were lower than the prediction of Equation (6).

$$\Delta T_{\mathrm{m}} = \frac{\dot{Q}}{V b_{fo}^{1/3} H_{ef}^{5/3}}, \quad V' > 0.19 \tag{6}$$

$$\begin{aligned} V' &= V'_{C+D} = \frac{V_{C+D}}{w^*} \\ w^* &= \left(\frac{\dot{Q} g}{b_{fo} \rho_0 c_p T_0} \right)^{1/3} \end{aligned} \tag{7}$$

where:

V_{C+D} is the sum of ramps C and D velocity (m/s).

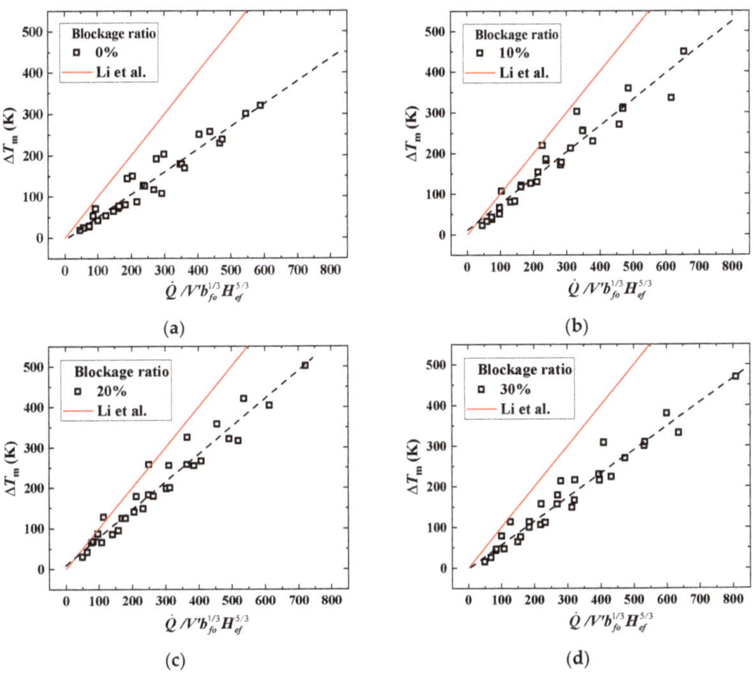

Figure 10. Fitting of maximum temperature rise with different blockage ratios: (**a**) ϕ = 0%; (**b**) ϕ = 10%; (**c**) ϕ = 20%; (**d**) ϕ = 30%.

This is because the maximum temperature rise prediction equation proposed by Li is based on ventilation tests in the single-pipe tunnel. The maximum temperature rise in this paper was based on the interconnected tunnel. At the same velocity, the tunnel section where the fire source was located was larger than the tunnel section in the Li model, so the air volume was also larger than that in the Li model. Ramp C and ramp D supplied air at the same time, which increased the contact area between the fire plume and the airflow, so the air entrainment of the fire plume in the tunnel section also increased, which resulted in the decrease in the temperature. Therefore, the experiment data were lower than Equation (6).

Figure 11 contrasts the maximum temperature increases with various blockage ratios with the predictions of the Li model. It was discovered that there were no obvious fluctuation tendencies in the distribution under the various blockage ratios. This demonstrates that when the interconnected tunnel was blocked, Equation (6) was not capable of forecasting the increase in temperature.

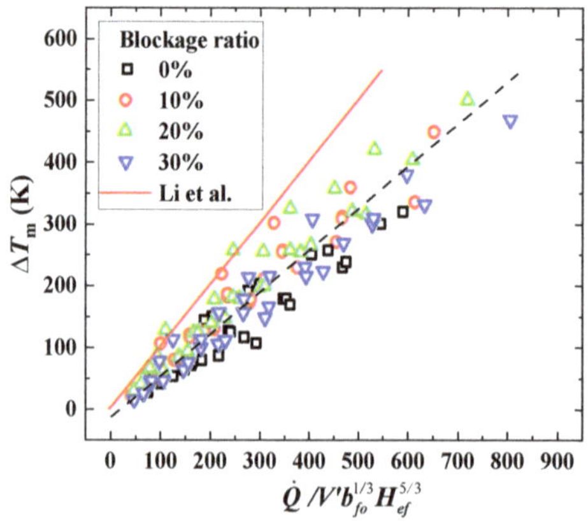

Figure 11. Comparison of the maximum temperature rise with the prediction equation of Li.

Equation (6) focuses on the maximum temperature rise of the single-pipe tunnel. To better understand the maximum temperature rise during the interconnected tunnel's blockage, this paper establishes a suitable maximum temperature rise model for different blockage ratios of the test data. The effects of vehicle blockage, longitudinal ventilation velocity, and HRR on the maximum temperature increase in an interconnected tunnel were studied in this model, where the ventilation velocity of ramp D was directly affected by the vehicle blockage.

In the underground interconnected tunnel, the maximum temperature rise was affected by the velocities of both ramp C and ramp D, while the velocity of ramp D changed with the vehicle blockage ratios. The longitudinal velocity and blocking ratio were combined with interconnected tunnels, and this paper analyzed the smoke mass flow rate to establish a prediction model.

The smoke mass flow was mainly composed of the fuel mass flow and air mass flow in ramp C and ramp D.

$$\dot{m}_s = \dot{m}_C + \dot{m}_D + \dot{m}_f \tag{8}$$

The fuel mass flow rate was much lower than that of ramp C and ramp D. Therefore, the mass flow rate of smoke can be expressed by the sums of the longitudinal ventilation velocities of ramps C and D:

$$\dot{m}_s = \rho V_{C+D} S \tag{9}$$

where:

S is the tunnel cross-sectional area (m^2).

Based on the heat calculation formula, where $\dot{Q}_c = 0.7\dot{Q}$:

$$\dot{Q}_c = C \times \dot{m} \times \Delta T_m \tag{10}$$

where C is a constant.

The maximum temperature rise is proportional to $\dot{Q}_c/\dot{m}$, as shown in Equation (11):

$$\Delta T_m \propto \frac{\dot{Q}_c}{\rho V_{C+D} S} \tag{11}$$

When the blockage ratio is 0%, the relation (12) is obtained from Equation (11), which is shown in Figure 12 by fitting the data.

$$\Delta T_{m,0\,\%} \propto \frac{\dot{Q}_c}{\rho v_{C+D} S} \tag{12}$$

where:

v_{C+D} is the sum of the velocity of ramps C and D when the blockage ratio is 0%.

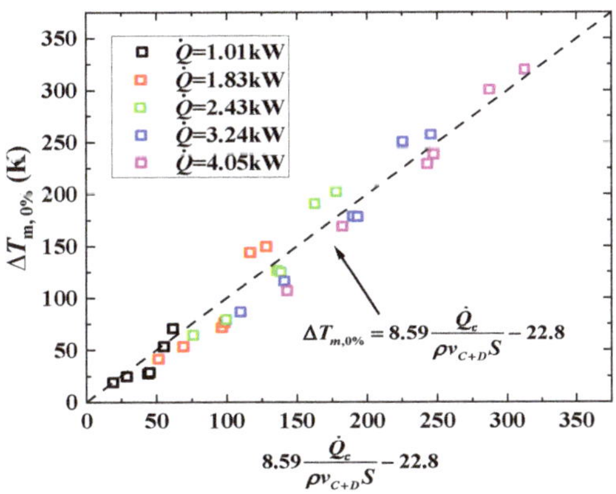

Figure 12. Maximum temperature rise prediction formula (ϕ = 0%).

As can be seen from Figure 12, the prediction equation at 0% blockage is shown in Equation (13), which agrees well with the experimental results.

$$\Delta T_{m,0\,\%} = 8.59\frac{\dot{Q}_c}{\rho v_{C+D} S} - 22.8,\ \ 0.6 \le v_{C+D} \le 1.6 \tag{13}$$

For the velocity sum of ramps C and D, introducing the blockage ratio ϕ, the following relationship can be obtained:

$$V_{C+D} = v_{C+D}\, g(\phi) \tag{14}$$

As the blockage ratio increases in the interconnected tunnel, the sum of the velocities at ramps C and D becomes lower. If the relationship between V_{C+D} and v_{C+D} satisfies Equation (15), the data are fitted by fitting the data as shown in Figure 13.

$$g(\phi) = V_{C+D}/v_{C+D} = (1 - a\phi) \tag{15}$$

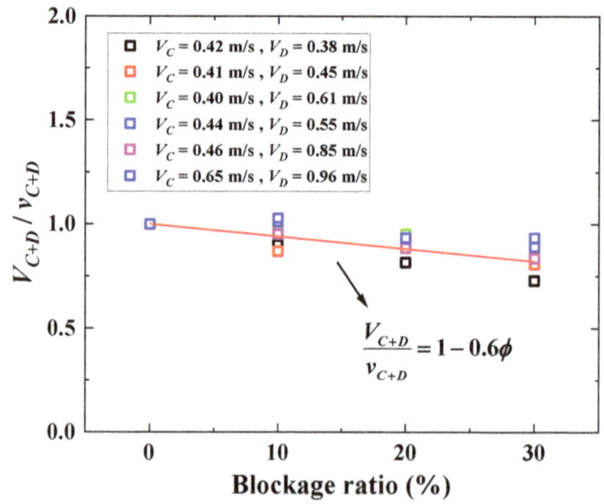

Figure 13. Variation in velocities with blockage ratio.

The relationship between the ratio V_{C+D} and v_{C+D} and the blockage ratio can be obtained from Figure 13, as shown in Equation (16).

$$g(\phi) = V_{C+D}/v_{C+D} = (1 - 0.6\phi),\ \ 0\% \leq \phi \leq 30\% \tag{16}$$

Equations (13) and (16) may be used to derive the link (11) between the maximum temperature rise and velocity and blockage ratio, and the data are fitted, as shown in Figure 14. The test data under different blockage ratios of the interconnected tunnel fit well with the prediction formula, while Li's [22] formula cannot predict it well, as shown in Figure 15. The prediction formula is shown in Equation (17). This formula applies to interconnected tunnel blockage rates between 0% and 30%.

$$\Delta T_m = 8.59\frac{\dot{Q}_c}{\rho v_{C+D}(1 - 0.6\phi)S} - 22.8,\ \ 0.6 \leq v_{C+D} \leq 1.6 \tag{17}$$

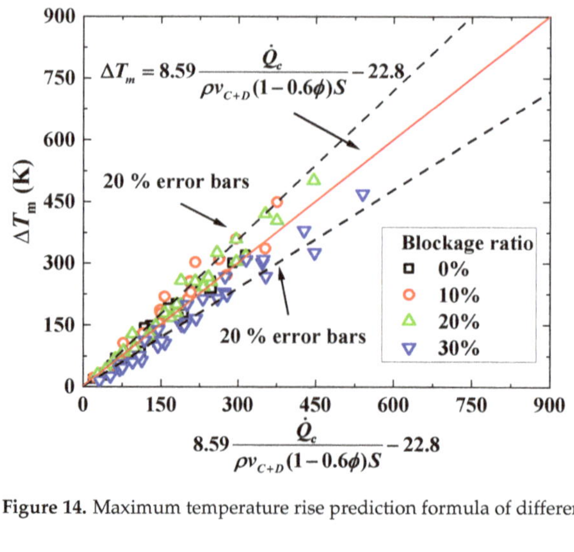

Figure 14. Maximum temperature rise prediction formula of different blockage ratios.

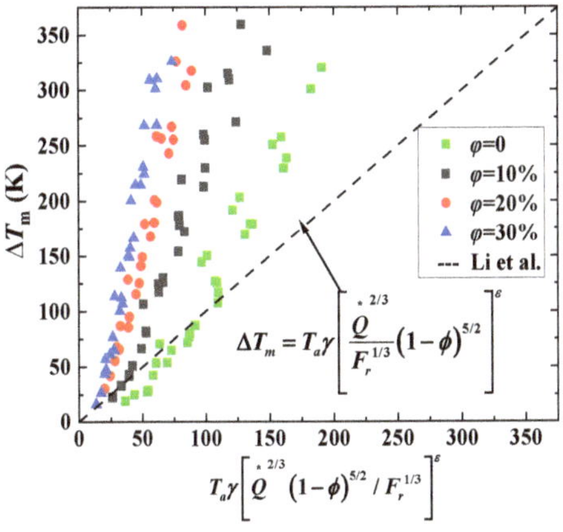

Figure 15. Comparison between the Li's model and the experimental data considered in this study.

4. Conclusions

The model-scale tunnel calculates the maximum temperature increase in the smoke under the interconnected tunnel by combining the velocity and HRR of the fire source under various blockage ratios. The following are the key conclusions of this paper:

(1) The velocity of the ramps upstream of the fire source and the adjacent ramp have an impact on the maximum temperature rise in the interconnected tunnel. The maximum temperature rise is jointly impacted by both ramps' ventilation.

(2) The maximum temperature rise in the interconnected tunnel varies with velocity and the blockage ratio. Depending on who is more impacted by the velocity or blockage ratio, the maximum temperature rise differs. The maximum temperature rise decreases, and the impact of the blockage ratio diminishes when the velocity in the interconnected tunnel decreases. This is because the convective heat transfer near the fire source increases.

(3) In an underground interconnected tunnel fire, the maximum temperature rise is influenced by the velocity upstream of the fire source and the adjacent tunnel. As a result, this research provides a novel method for predicting the maximum temperature rise under the underground interconnected tunnel that is blocked by vehicles.

We only studied blockage rates between 0% and 30% in this paper. Additionally, the angles between ramp C and ramp D were not considered. More research will be carried out in the next stage.

Author Contributions: Methodology, Z.X.; software, J.Z.; validation, H.Y.; formal analysis, H.Y.; investigation, H.Y.; resources, B.X.; data curation, Y.Z.; writing—original draft preparation, Y.Z.; writing—review and editing, S.M.S.T. All authors have read and agreed to the published version of the manuscript.

Funding: The Science and Technology Research and Development Program Project of China railway group limited (Major Special Project, No.: 2021-Special-04-2) funded this research.

Institutional Review Board Statement: Not applicable.

Informed Consent Statement: Not applicable.

Data Availability Statement: Not applicable.

Acknowledgments: We acknowledge the High-Performance Computing Center of Central South University for its support.

Conflicts of Interest: The authors declare no conflict of interest.

References

1. Guo, X.X.; Pan, X.H.; Zhang, L.J.; Wang, Z.L.; Hua, M.; Jiang, C.C. Comparative study on ventilation and smoke extraction systems of different super-long river-crossing subway tunnels under fire scenarios. *Tunn. Undergr. Space Technol.* **2021**, *113*, 103849. [CrossRef]
2. Zhou, T.N.; He, Y.P.; Lin, X.; Wang, X.H.; Wang, J. Influence of constraint effect of sidewall on maximum smoke temperature distribution under a tunnel ceiling. *Appl. Therm. Eng.* **2017**, *112*, 932–941. [CrossRef]
3. Kang, K. Characteristic length scale of critical ventilation velocity in tunnel smoke control. *Tunn. Undergr. Space Technol.* **2009**, *25*, 205–211. [CrossRef]
4. Alper, R.L. Calculation of response time of ceiling-mounted fire detectors. *Fire Technol.* **1972**, *8*, 181–195. [CrossRef]
5. Heskestad, G.; Delichatsios, M.A. The initial convective flow in fire. *Symp. Int. Combust.* **1979**, *17*, 1113–1123. [CrossRef]
6. Yi, L.; Xu, Q.Q.; Xu, Z.S.; Wu, D.X. An experimental study on critical velocity in sloping tunnel with longitudinal ventilation under fire. *Tunn. Undergr. Space Technol.* **2014**, *43*, 198–203. [CrossRef]
7. Zhang, X.L.; Lin, Y.J.; Shi, C.L.; Zhang, J.P. Numerical simulation on the maximum temperature and smoke back-layering length in a tilted tunnel under natural ventilation. *Tunn. Undergr. Space Technol.* **2021**, *107*, 103661. [CrossRef]
8. Hu, L.H.; Chen, L.F.; Wu, L.; Li, Y.F.; Zhang, J.Y.; Meng, N. An experimental investigation and correlation on buoyant gas temperature below ceiling in a slopping tunnel fire. *Appl. Therm. Eng.* **2013**, *51*, 246–254. [CrossRef]
9. Tang, F.; Cao, Z.L.; Palacios, A.; Wang, Q. A study on the maximum temperature of ceiling jet induced by rectangular source fires in a tunnel using ceiling smoke extraction. *Int. J. Therm. Sci.* **2018**, *127*, 329–334. [CrossRef]
10. Zhu, Y.T.; Tang, F.; Zhao, Z.X.; Wang, Q. Effect of lateral smoke extraction on transverse temperature distribution and smoke maximum temperature under ceiling in tunnel fires. *J. Therm. Anal. Calorim.* **2022**, *147*, 4275–4284. [CrossRef]
11. Li, Y.Z.; Ingason, H. Overview of research on fire safety in underground road and railway tunnels. *Tunn. Undergr. Space Technol.* **2018**, *81*, 568–589. [CrossRef]
12. Zhuang, Y.Z.; Ding, H.; Zheng, G.P.; Cui, Y.K.; Huang, Y. Study on Ventilation System Linkage Control Strategy in a Double-Hole Tunnel Fire. *Adv. Mater. Sci. Eng.* **2020**, *2020*, 5163632. [CrossRef]
13. Han, J.Q.; Liu, F.; Wang, F.; Weng, M.C.; Liao, S.J. Full-scale experimental investigation on smoke spreading and thermal characteristic in a transversely ventilated urban traffic link tunnel. *Int. J. Therm. Sci.* **2021**, *170*, 107130. [CrossRef]
14. Zhong, W.; Liu, L.; Han, N.; Gao, Z.H.; Zhao, J. Investigation on the maximum ceiling temperature of the weak plume impingement flow in tunnel fires under longitudinal ventilation. *Tunn. Undergr. Space Technol.* **2023**, *131*, 104821. [CrossRef]
15. Kurioka, H.; Oka, Y.; Satoh, H.; Sugawa, O. Fire properties in near field of square fire source with longitudinal ventilation in tunnels. *Fire Saf. J.* **2003**, *38*, 319–340. [CrossRef]
16. Li, Y.Z.; Lei, B.; Ingason, H. The maximum temperature of buoyancy-driven smoke flow beneath the ceiling in tunnel fires. *Fire Saf. J.* **2011**, *46*, 204–210. [CrossRef]
17. Li, L.M.; Cheng, X.D.; Cui, Y.; Li, S.; Zhang, H.P. Effect of blockage ratio on critical velocity in tunnel fires. *J. Fire Sci.* **2012**, *30*, 413–427. [CrossRef]
18. Luo, J.L.; Xu, Z.S.; Li, F.L.; Zhao, J.M. Effect of vehicular blocking scene on smoke spread in the longitudinal ventilated tunnel fire. *Case Stud. Therm. Eng.* **2019**, *14*, 100495. [CrossRef]
19. Tang, F.; Cao, Z.L.; Chen, Q.; Meng, N.; Wang, Q.; Fan, C.G. Effect of blockage-heat source distance on maximum temperature of buoyancy-induced smoke flow beneath ceiling in a longitudinal ventilated tunnel. *Int. J. Heat Mass Transf.* **2017**, *109*, 683–688. [CrossRef]
20. Meng, N.; Liu, B.B.; Li, X.; Jin, X.N.; Huang, Y.J.; Wang, Q. Effect of blockage-induced near wake flow on fire properties in a longitudinally ventilated tunnel. *Int. J. Therm. Sci.* **2018**, *134*, 1–12. [CrossRef]
21. Kayili, S.; Yozgatligil, A.; CahitEralp, O. An experimental study on the effects of blockage ratio and ventilation velocity on the heat release rate of tunnel fires. *J. Fire Sci.* **2011**, *29*, 555–575. [CrossRef]
22. Li, L.M.; Cheng, X.D.; Cui, Y.; Dong, W.H. Effect of blockage ratio on the maximum temperature under the ceiling in tunnel fires. *J. Fire Sci.* **2013**, *31*, 245–257. [CrossRef]
23. Huang, Y.B.; Li, Y.F.; Li, J.F.; Li, J.X.; Wu, K.; Zhu, K.; Li, H.H. Experimental investigation on maximum gas temperature beneath the ceiling in a branched tunnel fire. *Int. J. Therm. Sci.* **2019**, *145*, 105997. [CrossRef]
24. Chen, L.F.; Mao, P.F.; Zhang, Y.C.; Xing, S.S.; Li, T. Experimental study on smoke characteristics of bifurcated tunnel fire. *Tunn. Undergr. Space Technol.* **2020**, *98*, 103295. [CrossRef]

fire

Article

Numerical Simulation on the Effect of Fire Shutter Descending Height on Smoke Extraction Efficiency in a Large Atrium

Qiyu Liu [1], Jianren Xiao [2], Bihe Cai [3], Xiaoying Guo [2], Hui Wang [4], Jian Chen [5], Meihong Zhang [6], Huasheng Qiu [7], Chunlin Zheng [8] and Yang Zhou [1,*]

1 School of Civil Engineering, Central South University, Changsha 410075, China; liuqy99@csu.edu.cn
2 Fujian Provincial Institute of Architectural Design and Research Co., Ltd., Fuzhou 350001, China; xjr@fjadi.com.cn (J.X.); gxy@fjadi.com.cn (X.G.)
3 YanGuo Technology Co., Ltd., Xiamen 361001, China; bihe.cai@csu.edu.cn
4 Fujian Construction Engineering Group Co., Ltd., Fuzhou 350001, China; huiwang490@csu.edu.cn
5 China Academy of Building Research, Beijing 100013, China; ken_chenjian@126.com
6 Xiamen Fire Rescue Detachment, Xiamen 361013, China; zhangmhxm@sina.com
7 China Construction Fourth Engineering Division Co., Ltd., Guangzhou 510665, China; hsqiu1979@csu.edu.cn
8 Straits Construction Group Co., Ltd., Fuzhou 350014, China; clzheng896@csu.edu.cn
* Correspondence: zyzhou@csu.edu.cn

Abstract: In this study, a series of numerical simulations were carried out to investigate the effect of fire shutter descending height on the smoke extraction efficiency in a large space atrium. Based on the full-scale fire experiments, this paper carried out more numerical simulations to explore factors affecting the smoke extraction efficiency in the atrium. The smoke flow characteristics, temperature distribution law and smoke extraction efficiency of natural and mechanical smoke exhaust systems were discussed under different heat release rates and fire shutter descending heights. The results show that the smoke spread rate and the average temperature of the smoke are higher with a greater heat release rate. After the mechanical smoke exhaust system is activated, the smoke layer thickness and smoke temperature decrease, and the stable period of heat release rate is shorter. In the condition of natural smoke exhaust, the smoke extraction efficiency increases exponentially with the increase of heat release rate and the descending height of the fire shutter, and the maximum smoke extraction efficiency is 48.8%. In the condition of mechanical smoke exhaust, the smoke extraction efficiency increases with the increase of mechanical exhaust velocity. When the velocity increases to the critical value (8 m/s), the smoke extraction efficiency is essentially stable. The smoke extraction efficiency is increased first with the increase of fire shutter descending height and then has a downward trend when the descending height drops to half, and the maximum smoke extraction efficiency is 70.3% in the condition of mechanical smoke exhaust. Empirical correlations between the smoke extraction efficiency and the dimensionless fire shutter descending height, the dimensionless heat release rate and the dimensionless smoke exhaust velocity have been established. The results of this study can provide a reference for the design of smoke prevention and exhaust systems in the atrium.

Keywords: atrium fire; smoke extraction efficiency; fire shutter descending height; numerical simulation

Citation: Liu, Q.; Xiao, J.; Cai, B.; Guo, X.; Wang, H.; Chen, J.; Zhang, M.; Qiu, H.; Zheng, C.; Zhou, Y. Numerical Simulation on the Effect of Fire Shutter Descending Height on Smoke Extraction Efficiency in a Large Atrium. *Fire* **2022**, *5*, 101. https://doi.org/10.3390/fire5040101

Academic Editors: Chuangang Fan and Dahai Qi

Received: 14 June 2022
Accepted: 15 July 2022
Published: 17 July 2022

1. Introduction

In recent years, an increasing number of urban commercial complexes have emerged, not only bringing great convenience to people's lives, but also causing greater fire hazards. The mall atrium is a large space which integrates shopping, entertainment, leisure and other functions. Usually, there are dense personnel and numerous combustibles in the mall atrium, which brings a high fire risk. Once a fire occurs, it is likely to spread to the buildings next to the atrium, which causes serious economic losses and casualties; moreover, smoke is the most critical factor that threatens the safety of people in fires. Therefore, it is of great significance to study the law of smoke spreading and smoke extraction efficiency in the atrium and indoor pedestrian street, which is the key passage for personnel safety evacuation [1–3].

The research on the spread and control of fire smoke in large-space buildings started in the 1970s. Many scholars have studied the factors affecting the fire plume and smoke spread characteristics, and given many mathematical relationships through experiments, such as Thomas et al. [4], McCaffery [5,6], Zukoski et al. [7,8], Heskestad [9,10] and Tanaka et al. [11]. In addition, they established some theoretical models of axisymmetric fire plumes, which laid the theoretical foundation for the study of fire smoke development in large spaces. Since then, their results have been widely used in the prediction and control of fire smoke. With the increase of research by domestic and foreign scholars, the understanding of the fire smoke spread characteristics has gradually increased and a relatively complete system has been formed. In the 1990s, with the development of computational technology, the advantages of numerical simulation technology have gradually been highlighted, which has good visualization effects and simple operation methods. Nowadays, it has become the mainstre am method of studying indoor large-space fire [12–16].

Fires generate a great deal of heat and smoke, which threaten people's lives. On the one hand, fire and smoke separations are required to hinder the further spread of fire and smoke; on the other hand, heat and smoke also need to be exhausted from the atrium by smoke vents. Therefore, the efficiency of smoke extraction is closely linked to the safe evacuation of people. Different smoke exhaust systems will cause different smoke flowing and temperature distribution laws. As a rule, mechanical smoke exhaust systems are more responsive and reliable than natural smoke exhaust systems, which are now widely used for fire and smoke extraction in large spaces. Many scholars have carried out exhaustive research on large space atrium fires. For example, Klote et al. [17,18] found the law and calculation method of smoke flow in atrium fire. Hadjisophocleous et al. [19–22] combined experiments and numerical simulation to analyze the smoke spread and filling characteristics of atrium fires, the calculation method of smoke layer height and the smoke extraction efficiency of the mechanical smoke exhaust system in the atrium. Rho and Ryou [23] studied the characteristics of smoke flow in a large space atrium by numerical simulation. Chow et al. [24–26] investigated the fire development form and the smoke flow characteristics of the atrium through numerical simulation and small-scale experiments. Wong [27] figured out the influence of the distance between the fire shutter and the evacuation distance of the emergency exit under various circumstances. Yu and Wei [28] ensured the safe distance from combustibles to the fire shutters in the atrium. Yu [29] aimed at the safety and reliability of the fire shutter.

Long et al. [30] carried out full-scale fire experiments for several vital parameters including the vertical temperature distribution, the longitudinal temperature distribution, smoke layer height and smoke front arrival time under four different scenarios. Huang et al. [31] found the enlargement of the difference between the two heights, and the smoke spread process became slower at the constant heat release rate. Zhang et al. [32] solved the impact of the segmented smoke exhaust of the ultra-thin and tall atrium on fire prevention and control by using a full-scale hot smoke experiment method. Xu's results via the fire dynamics simulator (FDS) indicate that the temperature is changed more sharply than the visibility while the ceiling height gets higher [16].

Although scholars have studied the smoke extraction efficiency of the atrium through experiments or simulations, there is little involved in the impact of the fire shutter on it. The existence of a fire shutter has significantly changed the fire situation; this results in a research gap.

A fire shutter is a facility used for fire protection, heat insulation and smoke prevention for buildings; it also has an effect on the efficiency of smoke extraction in large spaces. The descending height of the fire shutter will affect the airflow, thus affecting the smoke extraction efficiency; besides, the smoke extraction method (including natural smoke extraction and mechanical smoke extraction) also has an effect on the smoke extraction efficiency; however, current studies have not given the influence mechanism of fire shutter on the smoke extraction effect. Therefore, this study is expected to reveal the relationship between the two parameters by numerical simulations.

2. Numerical Simulation Conditions

2.1. Governing Equations

The flow of smoke in a building fire exists in the form of turbulent flow. Several conservation laws should be followed in the process of turbulent flow, including conservation of mass, conservation of momentum, conservation of energy and conservation of components. Therefore, a turbulence model should be established when studying building fire, and so should the computer simulations. Various conservation laws are expressed in the form of governing equations, which are introduced below [33].

(1) Equations for conservation of mass

$$\frac{\partial \rho}{\partial t}+\frac{\partial(\rho u)}{\partial x}+\frac{\partial(\rho v)}{\partial y}+\frac{\partial(\rho w)}{\partial z}=0 \quad (1)$$

where ρ is the air density, kg/m^3; t is time, s; u, v, w are the vectors in direction x, y, z.

(2) Equations for conservation of momentum

$$\frac{\partial(\rho u)}{\partial t}+\frac{\partial(\rho uu)}{\partial x}+\frac{\partial(\rho uv)}{\partial y}+\frac{\partial(\rho uw)}{\partial z}=\frac{\partial}{\partial x}\left(\mu\frac{\partial u}{\partial x}\right)+\frac{\partial}{\partial y}\left(\mu\frac{\partial u}{\partial y}\right)+\frac{\partial}{\partial z}\left(\mu\frac{\partial u}{\partial z}\right)-\frac{\partial p}{\partial x}+S_u \quad (2)$$

$$\frac{\partial(\rho v)}{\partial t}+\frac{\partial(\rho vu)}{\partial x}+\frac{\partial(\rho vv)}{\partial y}+\frac{\partial(\rho vw)}{\partial z}=\frac{\partial}{\partial x}\left(\mu\frac{\partial v}{\partial x}\right)+\frac{\partial}{\partial y}\left(\mu\frac{\partial v}{\partial y}\right)+\frac{\partial}{\partial z}\left(\mu\frac{\partial v}{\partial z}\right)-\frac{\partial p}{\partial y}+S_v \quad (3)$$

$$\frac{\partial(\rho w)}{\partial t}+\frac{\partial(\rho wu)}{\partial x}+\frac{\partial(\rho wv)}{\partial y}+\frac{\partial(\rho ww)}{\partial z}=\frac{\partial}{\partial x}\left(\mu\frac{\partial w}{\partial x}\right)+\frac{\partial}{\partial y}\left(\mu\frac{\partial w}{\partial y}\right)+\frac{\partial}{\partial z}\left(\mu\frac{\partial w}{\partial z}\right)-\frac{\partial p}{\partial z}+S_w \quad (4)$$

where ρ is the air density, kg/m^3; t is time, s; u, v, w are the vectors in direction x, y, z; p is the pressure of fluid microelement, Pa; S_u, S_v, S_w are generalized source term.

(3) Equations for conservation of energy

$$\frac{\partial(\rho T)}{\partial t}+\frac{\partial(\rho uT)}{\partial x}+\frac{\partial(\rho vT)}{\partial y}+\frac{\partial(\rho wT)}{\partial z}=\frac{\partial}{\partial x}\left(\frac{k}{c_p}\frac{\partial T}{\partial x}\right)+\frac{\partial}{\partial y}\left(\frac{k}{c_p}\frac{\partial T}{\partial y}\right)+\frac{\partial}{\partial z}\left(\frac{k}{c_p}\frac{\partial T}{\partial z}\right)+S_T \quad (5)$$

where T is temperature, K; k is heat transfer coefficient; c_p is specific heat capacity, kJ/(kg·K); S_T is viscous dissipation term.

(4) Equations for conservation of component

$$\frac{\partial(\rho c_S u)}{\partial x}+\frac{\partial(\rho c_S v)}{\partial y}+\frac{\partial(\rho c_S w)}{\partial z}=\frac{\partial}{\partial x}\left(D_s\frac{\partial \rho c_s}{\partial x}\right)+\frac{\partial}{\partial y}\left(D_s\frac{\partial \rho c_s}{\partial y}\right)+\frac{\partial}{\partial z}\left(D_s\frac{\partial \rho c_s}{\partial z}\right)+S_s \quad (6)$$

where c_s is the volume concentration of component S; D_S is the mass of the component S generated by the chemical reaction of unit time and unit volume

(5) Boundary condition

$$-k_s\frac{\partial T_s}{\partial t}=\dot{q}_c''+\dot{q}_r'' \quad (7)$$

where q_c'' is convective heat transfer, W; q_r'' is radiant heat, W.

2.2. Physical Model

This study takes the indoor pedestrian street of a children's hospital in Fujian Province as the model, which has 4 floors and the height is 4.5 m per floor. The pedestrian street is 154 m long and 17 m wide. The width of the ring corridor is 4 m and the height under the ceiling is 2.8 m. The pedestrian street has three atriums, #1, #2 and #3, and the height of each atrium is 19.8 m (including the height of smoke storage bins). The #2 atrium is selected as the study object, which is 32.4 m long and 8 m wide. The fire shutters separate the ring corridor from the atrium, which is 7.3 m wide, 31.8 m long and 2.8 m high. During the experiments, the doors and windows on both sides of the ring corridor are closed. The volume of the smoke exhaust fan in this study is 64,200 m^3/h as the actual experiment building is. The schematic diagram of the numerical model of the building complex is shown in Figure 1.

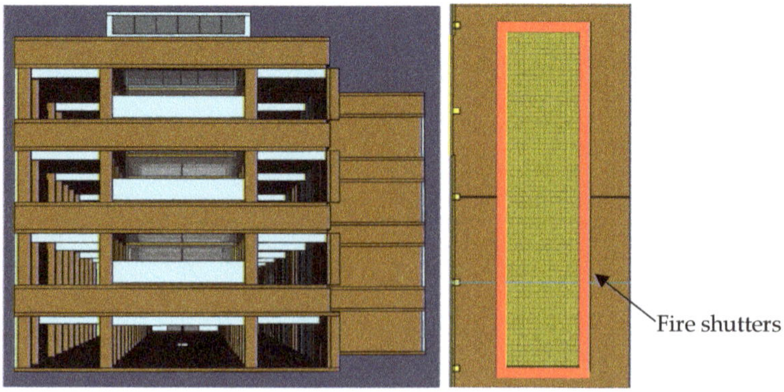

Figure 1. Schematic diagram of the numerical model of the building complex.

Fire usually occurs in the central location of the atrium floor, so the fire is set in the central location of the #2 atrium floor in this study. To simplify the calculation, the ground and structures are assumed to be adiabatic and the ambient temperature is 20 °C. The exterior walls are set as thermal insulation. There is no natural wind speed in the atrium. The atmospheric pressure is 101,325 Pa. In this study, ethanol is used as fuel to simulate the fire source with a t^2 fire model, which is assumed that the heat release rates keep constant after reaching the maximum. Besides, the temperature distribution and smoke extraction efficiency in the atrium are studied by numerical simulation under different fire shutter descending heights and exhaust conditions.

The fire shutter, the surface of which is made up of steel or inorganic fibre material, is a kind of fire prevention and heat insulation facility suitable for large openings in buildings. The fire shutters are driven through the transmission device and control system, playing a part of fire prevention and fire isolation; it is an indispensable fire prevention facility in modern buildings.

Heat release rate, smoke exhaust velocity, fire shutter descending height and exhaust method are selected as variables for numerical simulation. In total, 168 groups of experiments are designed. Three heat release rates are designed, 0.75 MW, 1.5 MW and 3.0 MW, respectively [34]. Eight smoke exhaust velocities are designed, ranging from 0 to 14 m/s with an interval of 2 m/s. The descending heights of the fire shutter are designed as 0 m, 0.7 m, 1.1 m, 1.4 m, 1.7 m, 2.1 m and 2.8 m, counting 7 groups.

Five thermocouples are arranged directly above the fire. The lowest thermocouple is 0.9 m away from the fire source, and the next is arranged every other 0.35 m upward, numbering 1–5 in sequence. The top thermocouple is 2.3 m away from the fire. More thermocouples are set from No. 5 thermocouples in the four directions, with an interval of 0.35 m, numbering 6–13 in turn. Thermocouples are arranged every other 1 m at the place 2.3 m away from the fire, and the highest one is 17.3 m, numbering V1-V16 from the top to the bottom. At the height of 9.3 m, 13.8 m and 18.3 m, 7 thermocouples are arranged along the centerline from west to east separately, with a horizontal distance interval of 1 m, numbering Z3F1-Z3F7, Z4F1-Z4F7, ZTH1-ZTH7. Temperature slices and visibility slices are set at the central line of ring corridors in every floor and the atrium centre. Several gas detectors are set near the window at the rooftop to monitor CO_2 concentration.

2.3. Grid-Dependency Evidence

In FDS numerical simulation, the grid distribution of the computational region must correctly describe the variation of physical quantities in the flow field, as well as meet the computer workload. Therefore, a sensitivity analysis of the grid is required. In the meshing of the model, the mesh edge length is decremented from 0.6 m to 0.3 m with an interval of 0.1 m each time, and every condition is run separately. The temperature variation of each working condition is compared with the experimental value of the same point, and the calculation accuracy increases gradually with the decrease of grid edge length. When the temperature of the measurement point is very close to the experimental value, the grid size at this time is the ideal simulation grid size. Figure 2 shows the temperature variation of the same measurement point with the grid side length from 0.6 m to 0.3 m with an interval of 0.1 m; it is easy to find that when the grid size is 0.3 m, the temperature is very close to the experimental value. If the grid size is further decreased, it will not help much to improve the accuracy, but the computation volume will increase four times. Therefore, the size of the grid for numerical simulation is determined to be 0.3 m × 0.3 m × 0.3 m in this study under the premise of ensuring both computational accuracy and computational efficiency. As is shown in Figure 2, the numerical simulation results are in good agreement with the experimental results and the data error is within an acceptable range [35]. Therefore, we believe that the calculated results of the numerical simulation in this study are reliable.

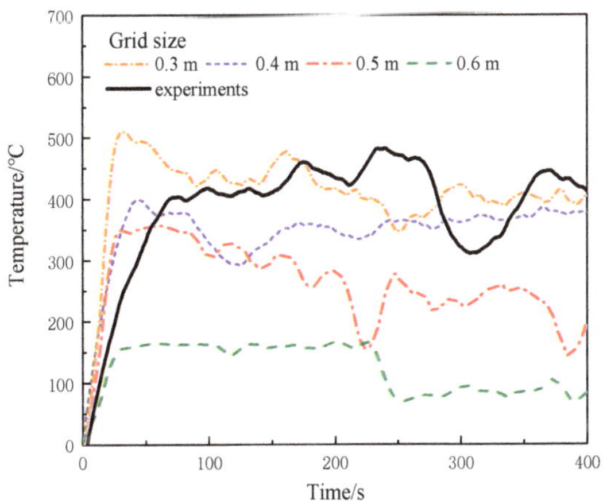

Figure 2. Temperature comparison chart for different grid sizes.

3. Results and Discussion

3.1. Smoke Spread Process

When a fire occurs, a large amount of smoke and heat will be generated to form a hot smoke flow, and the direction of smoke flow is often the direction of fire spread [36]. The smoke generated by the atrium fire will gradually converge over the atrium, and the smoke layer will continue to settle. Therefore, the visibility is constantly reduced and the visual range of evacuees is affected, slowing down the evacuation speed.

According to the principle of fire dynamics, fire development goes through three stages: accelerated combustion, stable combustion and extinction. Figure 3 shows the diagram of the smoke spread of fire inside the atrium at 30 s, 100 s and 300 s of the fire. Under the effect of thermal buoyancy, fire smoke rises first and spreads horizontally after reaching the ceiling. When the heat release rate is 0.75 MW, only a small amount of smoke is produced, and the height of smoke reaches half of the height of the atrium; after that, it begins to generate more smoke. At 100 s, the smoke reaches the ceiling and spreads into the ring corridor. At 300 s, the smoke has filled the upper space of the atrium. When the heat release rate reaches 1.5 MW, smoke is generated more obviously and quickly. The atrium above the fourth floor is almost filled with smoke at 100 s. When the heat release rate comes to 3.0 MW, smoke only takes 30 s to reach the ceiling, then it begins to spread outside through the window of the top floor at 100 s. Due to the restriction of the walls of the atrium, the continuously generated smoke begins to settle, and there is a clear boundary between the smoke layer and the cold air layer.

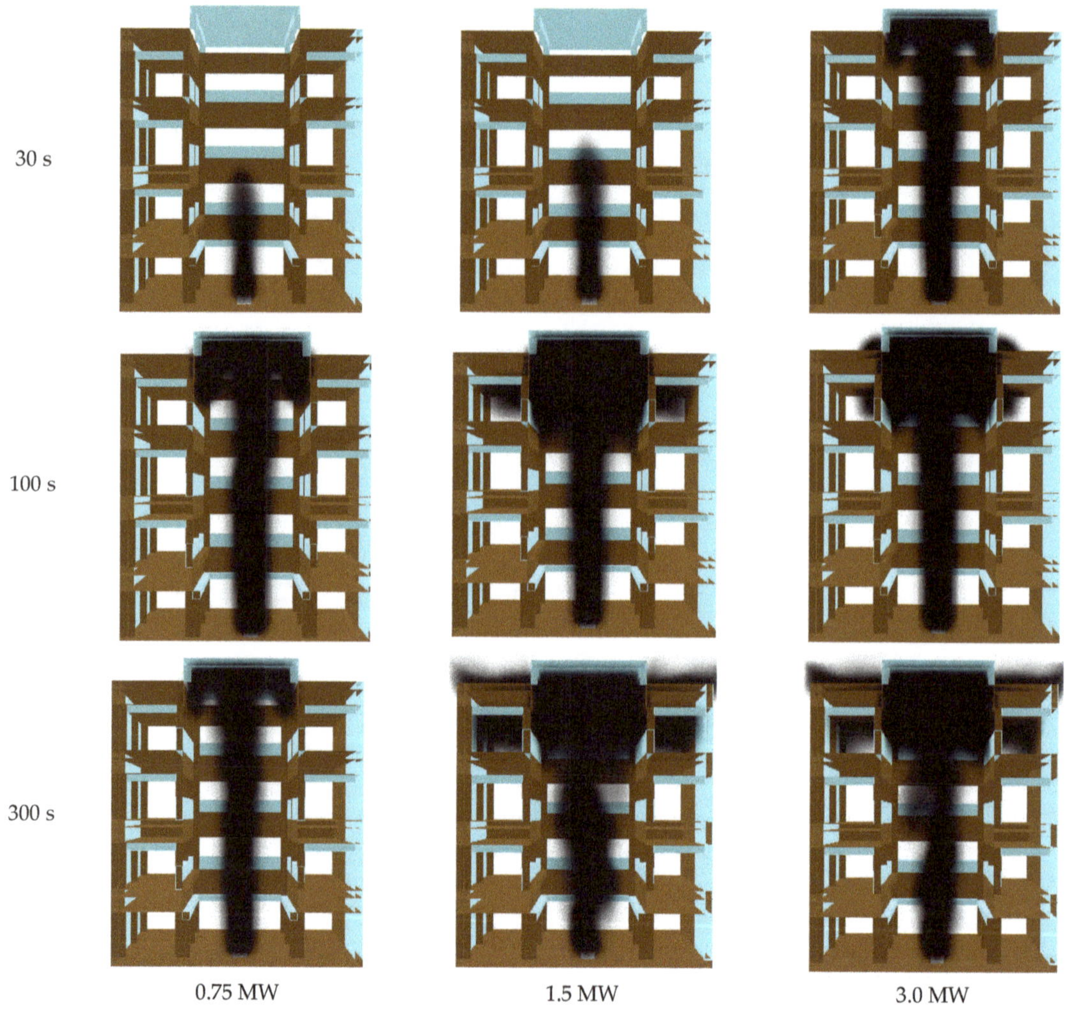

Figure 3. Diagram of the smoke spread of fire inside the atrium at different heat release rates and times.

There are significant differences in the rate of smoke spread under different heat release rates. In the case of a greater heat release rate, the smoke spread rate is faster, and the thickness of the stable smoke layer is significantly thicker. Obviously, the bigger the heat release rate is, the more obvious the heat buoyancy effect is, and the stronger the roll absorption effect is. Therefore, the smoke reaches the roof, gathers, and begins to settle earlier with a bigger heat release rate. When the natural smoke and mechanical smoke exhausts are turned on, a significant reduction in the concentration and thickness of the smoke layer is shown. Because the opening smoke vent draws the smoke outside, reducing the concentration and the thickness of the smoke layer.

As a separation between the atrium and the ring corridor, the fire shutter hinders the smoke from spreading to the ring corridor to some extent. The descending height of the fire shutter has a certain influence on the time and rate of smoke spreading into the ring corridor. When the fire shutter is not activated (descending height is 0 m), the smoke can freely enter the ring corridor.

Visibility is the furthest distance that one can see the target object in a given environment; it is greatly reduced due to massive suspended particles and harmful gases in the smoke, hindering the safe evacuation of people. The minimum visibility in a large space is 10 m in fire [37]. Figure 4 shows the diagram of the visibility inside the atrium at 30 s, 100 s and 300 s. Visibility tends to decrease from the fire source to the top of the atrium. At 30 s, the rate and volume of smoke generation is small, so only vertical orientation at the central axis is affected. Visibility near the fire decreased to less than 15 m. Gradually, the smoke trends to enter the atrium and ring corridor. Relatedly, visibility in the ring corridor begins to decrease. Because most of the smoke is concentrated in the upper layer of the atrium, the visibility of the fourth layer is greatly affected, decreasing to 12 m at 100 s. Smoke fills the ring corridor when the fire develops stable at 300 s; at this time, visibility of the fourth ring corridor decreases to about 6 m, and the lower floors are less affected. Since the smoke does not spread to the lower floors, the visibility keeps in a normal range.

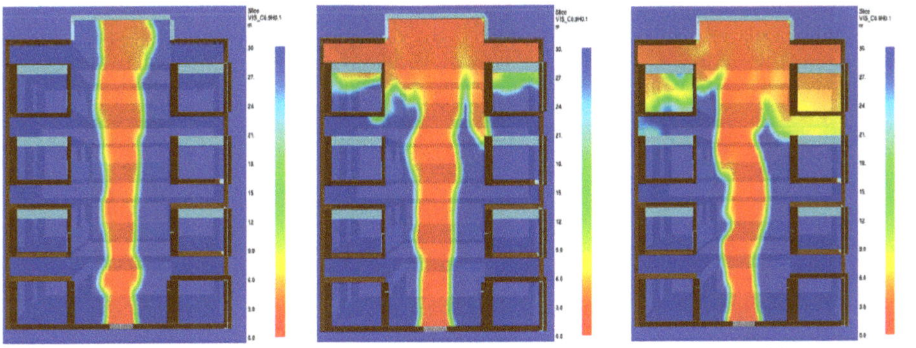

(a) (b) (c)

Figure 4. Diagram of the visibility inside the atrium at 30 s, 100 s and 300 s of the fire. (**a**) 30 s (**b**) 100 s (**c**) 300 s.

3.2. Temperature Distribution Law

A large amount of smoke and tremendous heat will be generated when a fire occurs. The hot environment and toxic smoke generated by the fire will make it difficult to evacuate the trapped people. Prolonged exposure to smoke can seriously damage a person's bodily functions and ability to escape.

Figure 5 shows the smoke temperature distribution inside the atrium at 30 s, 100 s and 300 s. At the early stage of the fire, less heat is released, so the smoke temperature rises relatively slowly. In the fire growth stage, the temperature gradient increases rapidly and reaches the maximum temperature quickly. With the decrease in heat release rate, the stable temperature value of the same measurement point nearest to the fire source decreases successively; it indicates that the greater the heat release rate is, the greater the average temperature of the smoke in the atrium is. The thermocouple closest to the fire source has the highest temperature, while the thermocouple farthest away has the lowest temperature. Because the smoke gradually loses heat as it moves farther up in the atrium. The maximum temperature at the centre of the fire source can reach about 750 °C. The natural smoke exhaust mainly affects the smoke temperature in the upper part of the atrium, and has less influence on the smoke temperature close to the fire source; this is mainly because only the top window of the atrium is opened during natural smoke exhaust, which accelerates the flow of smoke near it and accelerates the heat exchange; however, it has little effect on the near-fire source location, which is farther away from the glass window.

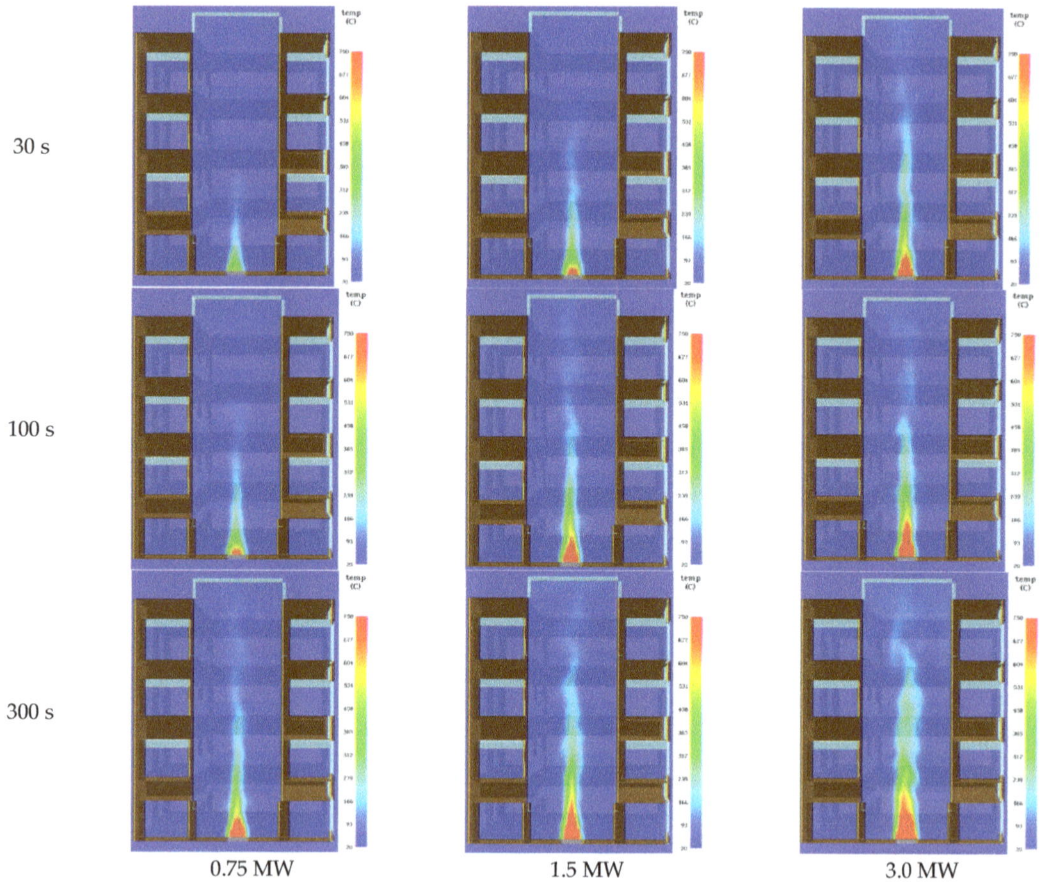

Figure 5. Fire smoke temperature distribution inside the atrium at different heat release rates and times.

3.3. Smoke Extraction Efficiency Analysis

The smoke extraction efficiency is defined as the percentage of smoke exhaust volume per unit time in the total amount of smoke generated, namely: smoke extraction efficiency = (smoke exhaust volume per unit time/the total amount of smoke generated per unit time) × 100% [38]; however, it is difficult to measure the amount of smoke produced and discharged because of the various components of smoke and the different entrainment amount of surrounding air by the plume under the influence of different external conditions. Therefore, CO_2, the main combustion product of ethanol, is selected as the calculation index in the numerical simulation. The ratio of CO_2 emission volume per unit time to CO_2

production volume per unit time is used to characterize the smoke extraction efficiency of the smoke extraction system. Smoke extraction efficiency δ can be described as:

$$\delta = \frac{CO_2\text{emission volume per unit time}}{CO_2\text{production volume per unit time}} \tag{8}$$

The chemical reaction equation for ethanol combustion in FDS is as follows [33]:

$$C_2H_5OH + 2.963O_2 \rightarrow 1.964CO_2 + 2.998H_2O + 0.04SOOT \tag{9}$$

The heat of combustion of ethanol ΔH = 27,000 kJ/kg. Therefore, it can be concluded that: CO_2 production volume per unit time is 0.055 kg/s in 0.75 MW fire, 0.11 kg/s in 1.5 MW fire, 0.22 kg/s in 3.0 MW fire.

In this paper, considering smoke extraction efficiency δ is related to the heat release rate $\dot{Q}$ and fire shutter descending heights h. A dimensionless analysis method is used to analyze the data.

Designing $\dot{Q}* = \frac{\dot{Q}}{\rho_\infty c_p T_\infty \sqrt{g} H^{5/2}}$, $h* = \frac{h}{H'}$, $v* = \frac{v}{v_{crit}}$ [39], where $\dot{Q}$ is total heat release rate; h is the descending height of the fire shutter; H' is the floor height; v is the smoke exhaust velocity; v_{crit} is the critical smoke exhaust velocity; ρ_∞ is the density of air around, 1.29 kg/m^3; c_p is the specific heat capacity of air at constant pressure, 1.005 kJ/(kg·K); T_∞ is the ambient temperature, 293 K; g is the acceleration of gravity, 9.8 N/kg. We can find $\delta \sim f\left(\dot{Q}*, h*, v*\right)$.

In the natural smoke exhaust mode, the relationship between the dimensionless descending height of the fire shutter and the smoke extraction efficiency is shown in Figure 6.

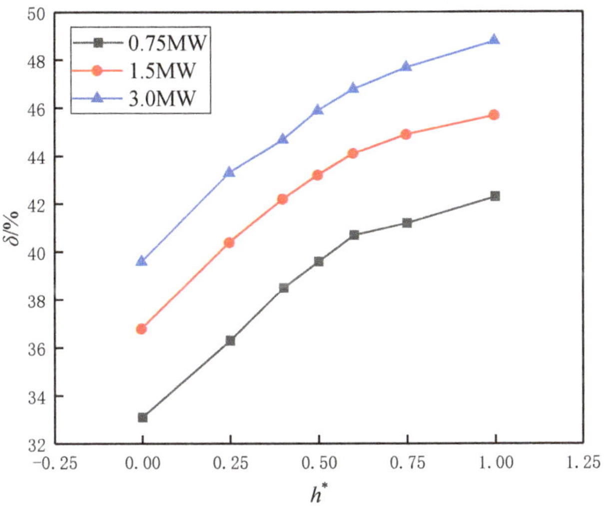

Figure 6. Relationship between the dimensionless descending height of the fire shutter and the smoke extraction efficiency.

In Figure 6, the smoke extraction efficiency increases steadily with the increase of fire shutter descending height in all conditions. In addition, the greater heat release rate possesses bigger smoke extraction efficiency, ranging from 39.6% to 48.8% at the biggest heat release rate while from 33.1% to 42.3% at the smallest heat release rate. When the heat release rate is 0.75 MW, the smoke extraction efficiency is only 33.1% if the fire shutter does not drop. The smoke extraction efficiency begins to increase along with the descending of the fire shutter, finally up to 42.3%. The initial growth is slightly faster than the latter. Trends of the smoke extraction efficiency when the heat release rates are 1.5 MW and 3.0 MW are similar to that of 0.75 MW, respectively, from 36.8% to 45.7% and from 39.6% to 48.8%. The increase of heat release rate enhances the average temperature of the smoke in the large space; this weakens the effect of air mixing into the smoke to some extent, so that the smoke extraction efficiency increases, and this phenomenon becomes more obvious when the heat release rate is greater.

As is shown above, $\dot{Q}* = \frac{\dot{Q}}{\rho_\infty c_p T_\infty \sqrt{g} H^{5/2}}$, $h* = \frac{h}{H}$, we can easily find that $\dot{Q}*$ and h^* are independent of each other. Considering δ is a function about $\dot{Q}*$ and h^*, we assume that $\delta \sim \left(\dot{Q}*\right)^m (h*)^n$. Fitting the δ with $\dot{Q}*$ and h^*, respectively, we can get $\delta \sim \left(\dot{Q}*\right)^{0.15}$, $\delta \sim (h*)^{0.1}$. The relationship between δ and $\left(\dot{Q}*\right)^{0.15}(h*)^{0.1}$ is shown in Figure 7.

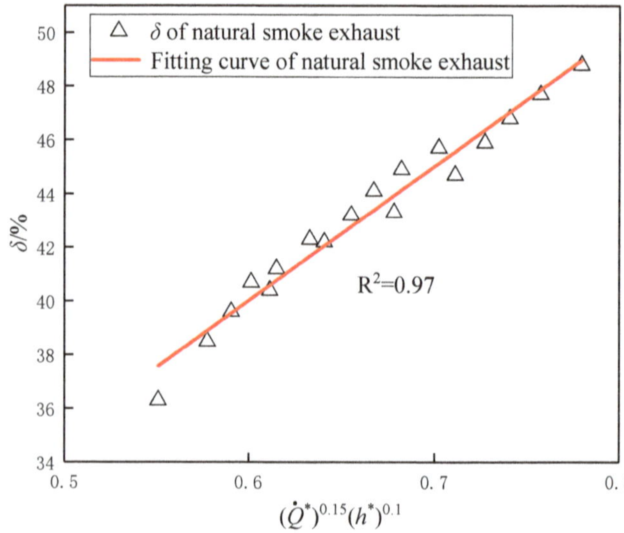

Figure 7. Natural smoke exhaust fitting line.

After linear fitting the data points above, we can find all the points fall almost near the same fit line. This line can be expressed as:

$$\delta = 10 + 50\left(\dot{Q}*\right)^{0.15}(h*)^{0.1} \tag{10}$$

Figure 7 shows a relationship between atrium smoke extraction efficiency and the dimensionless descending height of fire shutters and the dimensionless heat release rate. The points of abscissa are evenly distributed between 0.5 and 0.8. The natural smoke efficiency increases gradually from 36% to 48%. All the test points almost fall around the same line. The fitting coefficient R^2 reaches 0.97, which indicates a good fitting. Equation (10) can be empirically used to calculate the natural smoke extraction efficiency of the indoor pedestrian street atrium under the influence of a fire shutter.

Figure 8 shows the variation of smoke extraction efficiency δ with smoke exhaust velocity and descending height of fire shutters under different heat release rates.

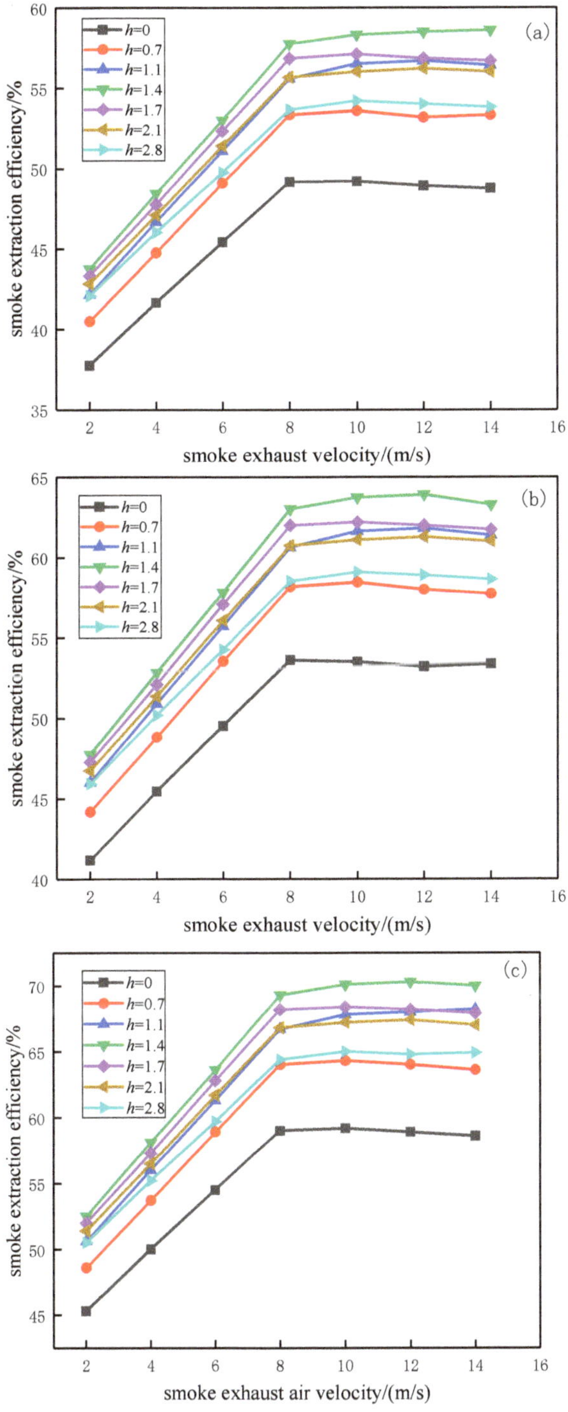

Figure 8. Curves of smoke extraction efficiency with smoke exhaust velocity and descending height of fire shutters at 0.75 MW, 1.5 MW, and 3.0 MW. (**a**) 0.75 MW (**b**) 1.5 MW (**c**) 3.0 MW.

When the smoke exhaust system is turned on, the air entrainment is intensified and the smoke rises with the air, so that the smoke extraction efficiency is gradually improved. When the fire shutter drops to a critical height of about 1.4 m (half of the floor height), the effect of the descending height on the smoke extraction efficiency is small, because the small space at the bottom restricts the air entrainment. As we can see, under different heat release rates, the smoke extraction efficiency at the critical height is stable at 58%, 64%, and 70%, respectively, with an interval of 6%.

In comparison, the efficiency of mechanical smoke exhaust is significantly higher than that of natural smoke exhaust. Because the smoke is continuously discharged from the atrium by the exhaust fan, which led to a better smoke extraction effect; thus, we suggest the atrium of the mall adopts the mode of combining natural smoke exhaust and mechanical smoke exhaust. Ordinarily, the natural smoke exhaust windows take the function of natural ventilation and smoke exhaust. Once a fire occurs, the mechanical smoke exhaust turns on and the fire shutters fall immediately. When the smoke exhaust air velocity stabilizes at 8 m/s and the fire shutters drop to half the floor height, the smoke extraction efficiency can reach the maximum.

It can be seen from Figure 8 that the smoke extraction efficiency in the mechanical smoke exhaust mode shows a trend of increasing at first with the increase of smoke exhaust velocity and then stabilizing after reaching a critical number. When the smoke exhaust velocity increases to 8 m/s, the mechanical smoke extraction efficiency is basically unchanged, regardless of the heat release rate. Therefore, we only analyze the increase stage of smoke extraction efficiency and ignore the stability stage. The dimensionless heat release rate $\dot{Q}*$ has no direct connection with dimensionless descending height h^* and dimensionless smoke exhaust velocity v^*. Therefore, it can be understood that $\dot{Q}*$ is independent with h^* and v^*. For the heat release rate of 3.0 MW, the variation law of δ with h^* and v^* are shown in Figures 9 and 10.

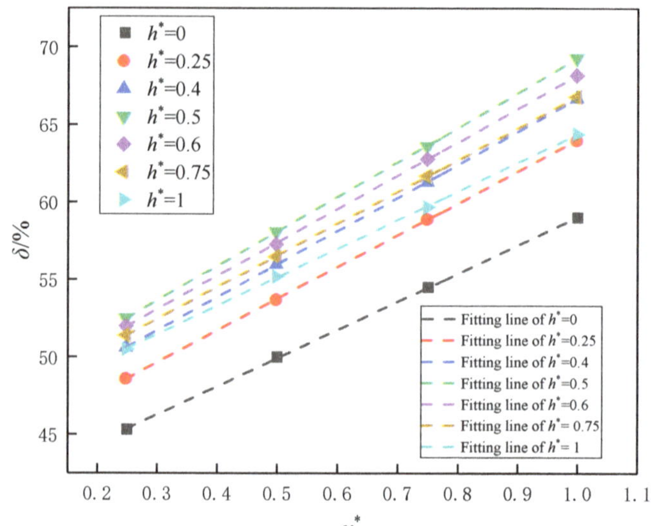

Figure 9. The relationship between δ and v^* when $\dot{Q}$ is 3.0 MW.

It can be seen from Figure 9 that δ and v^* show a linear positive correlation. The fitted linear relationship is shown in Table 1.

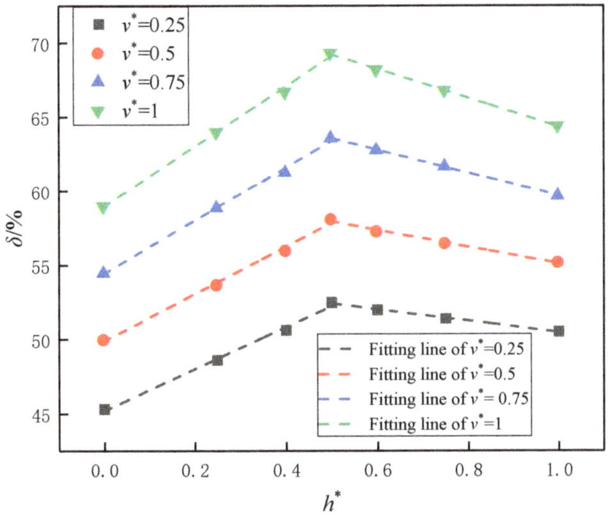

Figure 10. The relationship between δ and h^* when $\dot{Q}$ is 3.0 MW.

Table 1. Results between δ and v^*.

Equation				$\delta = a + bv^*$			
h^*	0	0.25	0.4	0.5	0.6	0.75	1
Intercept	40.8	43.5	45.3	46.9	46.6	46.3	45.9
Slope	18.3	20.6	21.4	22.4	21.6	20.6	18.5
R^2	0.99	0.99	0.99	0.99	0.99	0.99	0.99

The slope has good symmetry about the centre position and is related to dimensionless descending height h^*. This means v^* shows a symmetric function about δ; thus, the slope can be expressed as $8|2.8-(0.5-h*)|$. In other words, $\delta \sim f(h*)+8|2.8-(0.5-h*)|f(v*)$.

It can be seen from Figure 9 that δ and v^* show a symmetrical linear positive correlation; however, δ first linearly increases and then linearly decreases with the increase of h^*. Therefore, we describe the relationship between them in two parts, as Tables 2 and 3 shows.

Table 2. Results between δ and v^* (when $h \leq 1.4$ m ($h^* \leq 0.5$)).

Equation		$\delta = a + bh^*$		
v^*	0.25	0.5	0.75	1
Intercept	45.2	49.9	54.4	58.9
Slope	14.2	15.9	17.9	20.2
R^2	0.99	0.99	0.99	0.99

Table 3. Results between δ and v^* (when $h \geq 1.4$ m ($h^* \geq 0.5$)).

Equation		$\delta = a + bh^*$		
v^*	0.25	0.5	0.75	1
Intercept	54.4	60.8	67.5	70.1
Slope	−3.9	−5.7	−7.8	−9.7
R^2	0.99	0.99	0.99	0.99

When $h \leq 1.4$ m ($h^* \leq 0.5$), v^* shows a positive correlation with δ. The relation expression can be $\delta \sim [45.2 + 14.2h* + f(v*)]$; when $h \geq 1.4$ m ($h^* \geq 0.5$), v^* shows a negative correlation with δ. The relation expression can be $\delta \sim [52.5 - 4.3(h* - 0.5) + f(v*)]$. Therefore, $f(h*, v*)$ can be expressed as:

$$f(h*, v*) = \begin{cases} 45.2 + 14.2h* + 8(2.3 + h*)(v* - 0.25) & (h* \leq 0.5) \\ 52.5 - 4.3(h* - 0.5) + 8(3.3 - h*)(v* - 0.25) & (h* \geq 0.5) \end{cases} \quad (11)$$

Furthermore, take the heat release rate $\dot{Q}*$ into account, the relationship turns $\delta \sim \dot{Q}*^m f(h*, v*)$. The relationship between δ and $\dot{Q}*^m f(h*, v*)$ at different descending heights is shown in Figure 11. The fitting results show that the theoretical formula matches the actual results and is reliable.

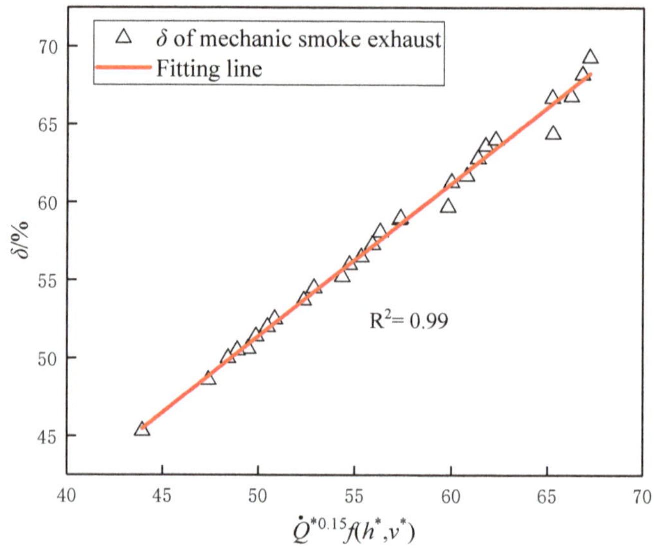

Figure 11. Relationship between δ and $\dot{Q}*^{0.15} f(h*, v*)$.

Therefore, when the mechanical smoke exhaust system is turned on, there is the following relationship between smoke extraction efficiency and smoke exhaust velocity and descending height of the fire shutter:

$$\delta = \dot{Q}*^{0.15} f(h*, v*) = \begin{cases} \dot{Q}*^{0.15}[45.2 + 14.2h* + 8(2.3 + h*)(v* - 0.25)] & (h* \leq 0.5) \\ \dot{Q}*^{0.15}[52.5 - 4.3(h* - 0.5) + 8(3.3 - h*)(v* - 0.25)] & (h* \geq 0.5) \end{cases} \quad (12)$$

4. Discussion and Conclusions

In this paper, the effect of the descending height of the fire shutter on the smoke spread law and smoke extraction efficiency in the atrium is studied by numerical simulation. When the fire heat release rate is small, the smoke subsidence speed in natural smoke exhaust mode is faster than that of mechanical smoke exhaust mode. At the same time, mechanical smoke exhaust can discharge a large amount of smoke and take away a large amount of heat in a timely manner. The mechanical smoke exhaust makes the temperature lower than that of natural smoke exhaust at the same time and measuring point.

In the large space of the shopping mall atrium, the natural smoke efficiency is lower than that of mechanical smoke exhaust. The efficiency of the natural smoke exhaust is about 40%, and the efficiency of the mechanical smoke exhaust is about 60%. Besides, some factors can also affect the smoke extraction efficiency of mechanical smoke exhaust, such as the heat release rate, the descending height of the fire shutter, and the smoke exhaust velocity.

After directional analysis, we draw some important conclusions as follows:

(1) With the increase of heat release rate from 0.75 MW to 3.0 MW, the smoke is generated and spreads faster. Besides the heat release rate being higher, the thickness of the stable smoke layer is significantly thicker than that with a lower heat release rate.

(2) When the heat release rate is lower than 1.5 MW, the subsidence rate of smoke in natural exhaust mode is faster than that in mechanical exhaust mode. The mechanical smoke exhaust system can discharge a large amount of smoke and heat in time, and the temperature at the same location is lower than using the natural smoke exhaust mode at the same time.

(3) In the natural smoke exhaust mode, the smoke extraction efficiency increases with the heat release rate and the descending height of the fire shutter, up to 48.8%. Empirical formula expression is expressed as Equation (12). The smoke flow is accelerated and the smoke extraction efficiency is increased, when the mechanical smoke exhaust system is turned on. The smoke extraction efficiency improves by 12% with the increase of the smoke exhaust velocity from 0 to 14 m/s. There is an upper limit value when the velocity is about 8 m/s, after which the smoke extraction efficiency is stable. Smoke extraction efficiency first increases with the increase of descending height. When the fire shutter down to half, the smoke extraction efficiency shows a downward trend. Smoke extraction efficiency reaches up to 70.3% after turning on the mechanical smoke extraction system. The empirical equation of smoke extraction efficiency has been established and the mechanical smoke extraction efficiency is about 20% higher than that of natural smoke extraction efficiency.

Although this paper takes a children's hospital as a model for experimental research, the conclusions can still be applied to other related structures, especially buildings with a large atrium and fire shutters. Whereas fire shutter plays a more and more important role in modern intelligent architecture, there is a lack of relative research on the influence of fire shutter on the smoke extraction effect. The results of this study can possess a guiding significance for the smoke control design of the architecture.

Author Contributions: Conceptualization, Q.L. and Y.Z.; methodology, Y.Z.; software, Q.L. and J.X.; validation, J.X., B.C. and Y.Z.; formal analysis, Q.L.; investigation, X.G.; resources, H.W.; data curation, J.C. and Q.L.; writing—original draft preparation, Q.L.; writing—review and editing, Q.L. and Y.Z.; visualization, B.C., X.G. and M.Z.; supervision, H.Q. and C.Z.; project administration, C.Z.; funding acquisition, J.X. and X.G. All authors have read and agreed to the published version of the manuscript.

Funding: This work was supported by the Regional Development Science and Technology Program of Fujian Province (No. 2019Y3002) and Natural Science Foundation of Hunan Province (No. 2021JJ30860).

Conflicts of Interest: The authors declare no conflict of interest.

References

1. Zheng, W. Fire safety design of indoor pedestrian streets of large commercial building. *Procedia Eng.* **2013**, *52*, 652–656. [CrossRef]
2. Xu, X.; Wang, Z.; Liu, X.; Ji, C.; Yu, N.; Zhu, H.; Li, J.; Wang, P. Study on Fire Smoke Control in Super-high Building Atrium. *Procedia Eng.* **2018**, *211*, 844–852. [CrossRef]
3. Wang, R.; Lan, X.; Xu, L. Smoke spread process under different heights based on numerical simulation. *Case Stud. Therm. Eng.* **2020**, *21*, 100710. [CrossRef]
4. Thomas, P.H.; Webster, C.T.; Raftery, M.M. Some experiments on buoyant diffusion Flames. *Combust. Flame* **1961**, *5*, 359–367. [CrossRef]
5. McCaffery, B.J. *Purely Buoyant Diffusion Flames: Some Experimental Results*; National Bureau of Standards: Washington, DC, USA, 1979.
6. McCaffery, B.J. Momentum implications for buoyant diffusion flames. *Combust. Flame* **1983**, *52*, 149–167. [CrossRef]
7. Zukoski, E.E. Development of a stratified ceiling layer in the early stage of closed room fire. *Fire Mater.* **1978**, *2*, 54–62. [CrossRef]
8. Zukoski, E.E.; Kubota, T.; Cetegen, B. Entrainment in fire plumes. *Fire Saf. J.* **1981**, *3*, 107–121. [CrossRef]
9. Heskestad, G. Virtual origins of fire plumes. *Fire Saf. J.* **1983**, *5*, 109–114. [CrossRef]

10. Heskestad, G. Engineering relations for fire plumes. *Fire Eng. J.* **1984**, *7*, 25–32. [CrossRef]
11. Tanaka, T.; Fujita, T.; Yamaguchi, J. Investigation into Rise Time of Buoyant Fire Plume Fronts. *Int. J. Eng. Perform. Based Fire Codes* **2000**, *2*, 14–25.
12. Guti´errez-Montes, C.; Sanmiguel-Rojas, E.; Viedma, A.; Rein, G. Experimental data and numerical modelling of 1.3 and 2.3 MW fires in a 20 m cubic atrium. *Build. Environ.* **2009**, *44*, 1827–1839. [CrossRef]
13. Qin, T.; Guo, Y.; Chan, C.; Lin, W. Numerical simulation of the spread of smoke in an atrium under fire scenario. *Build. Environ.* **2009**, *44*, 56–65. [CrossRef]
14. Ray, S.; Gong, N.; Glicksman, L.R.; Paradiso, J. Experimental characterization of full-scale naturally ventilated atrium and validation of CFD simulations. *Energy Build.* **2014**, *69*, 285–291. [CrossRef]
15. Zhang, G.; Zhou, X.; Zhu, G.; Yan, S. A new accident analysis and investigation model for the complex building fire using numerical reconstruction. *Case Stud. Therm. Eng.* **2019**, *14*, 100426. [CrossRef]
16. Xu, L.; Wang, Y.; Song, L. Numerical research on the smoke spread process of thin-tall atrium space under various ceiling height. *Case Stud. Therm. Eng.* **2021**, *25*, 100996. [CrossRef]
17. Klote, J.H. Prediction of smoke movement in atria: Part I—Physical concepts. *ASHRAE Trans.* **1997**, *103*, 534–544.
18. Klote, J.H. Prediction of smoke movement in atria: Part II—Application to smoke management. *ASHRAE Trans.* **1997**, *103*, 545–553.
19. Hadjisophocleous, G.V.; Lougheed, G.D.; Cao, S. Numerical study of the effectiveness of atrium smoke exhaust systems. *ASHRAE Trans.* **1999**, *105*, 1–17.
20. Lougheed, G.D. Investigation of atrium smoke exhaust effectiveness. *ASHARE Trans.* **1997**, *103*, 519–533.
21. Lougheed, G.D.; Hadjisophocleous, G.V.; Mecartney, C.; Taber, B.C. Largescale physical model studies for an atrium smoke exhaust system. *ASHRAE Trans.* **1999**, *105*, 676–698.
22. Rafinazari, A.; Hadjisophocleous, G.V. An investigation of the effect of make-up air velocity on smoke layer height with asymmetric openings and rotational air flow in atrium fires. *J. Build. Eng.* **2020**, *27*, 100933. [CrossRef]
23. Rho, J.S.; Ryou, H.S. A numerical study of atrium fires using deterministic models. *Fire Saf. J.* **1999**, *33*, 213–229. [CrossRef]
24. Chow, W.K. Smoke Movement and Design of Smoke Control In Atrium Building—Part2 Diagrams. *Int. J. Hous. Sci. Its Appl.* **1990**, *14*, 147–159.
25. Chow, W.K. Simulation of The Atrium Fire Environment in Hong Kong. *ASHRAE Trans.* **1993**, *99*, 163–168.
26. Chow, W.K. A Short Note on the Simulation of Atrium Smoke Filling Process Using Fire Zone Models. *J. Fire Sci.* **1994**, *12*, 516–528. [CrossRef]
27. Wong, L.T. Evaluation of Safe Distance of Fire rolling shutters in Shopping Malls. *Archit. Sci. Rev.* **2003**, *4*, 403–409. [CrossRef]
28. Yu, J.J.; Wei, G.Z. Distance from combustibles to fire rolling shutter of a large commercial building. *Fire Sci. Technol.* **2014**, *1*, 63–66.
29. Yu, H.C. The discussion of fire rolling shutter reliability. *Fire Sci. Technol.* **2014**, *6*, 701–703.
30. Long, Z.; Liu, C.; Yang, Y.; Qiu, P.; Tian, X.; Zhong, M. Full-scale experimental study on fire-induced smoke movement and control in an underground double-island subway station. *Tunn. Undergr. Space Technol.* **2020**, *103*, 103508. [CrossRef]
31. Huang, Y.; Wang, E.; Bie, Y. Simulation investigation on the smoke spread process in the large-space building with various height. *Case Stud. Therm. Eng.* **2020**, *18*, 100594. [CrossRef]
32. Zhang, H.; Wang, J.; Du, C. Experimental study on the effect of segmented smoke exhaust on smoke exhaust of ultra-thin and tall atrium. *Case Stud. Therm. Eng.* **2021**, *28*, 101560. [CrossRef]
33. Li, S.; Li, X. *FDS Fire Numerical Simulation*; Chemical Industry Press: Beijing, China, 2019.
34. *GA/T 999-2012*; Test Method for Verifying, Field Performance of Smoke Management System—Hot Smoke Test. The Ministry of Public Security of the People's Republic of China, Standards Press of China: Beijing, China, 2012.
35. Jiao, A.; Lin, W.; Cai, B.; Wang, H.; Chen, J.; Zhang, M.; Xiao, J.; Liu, Q.; Wang, F.; Fan, C. Full-scale experimental study on thermal smoke movement characteristics in an indoor pedestrian street. *Case Stud. Therm. Eng.* **2022**, *34*, 102029. [CrossRef]
36. Zhong, M.; Li, P.; Liu, T. Experimental study on fire smoke movement in a multi-floor and multi-room building. *Sci. China Ser. E Eng. Mater. Sci.* **2005**, *48*, 292–304. [CrossRef]
37. Hall, J.R. U.S. high-rise fires: The big picture. *NFPA J.* **1994**, *88*, 47–53.
38. Pan, Y.; Zhao, H.; Wu, D.; Li, W. Study on smoke exhaust efficiency under central exhaust mode in tunnel fires. *J. Saf. Environ.* **2012**, *12*, 191–196.
39. Hurley, M.J.; Gottuk, D.; Hall, J.R.; Harada, K.; Kuligowski, E.D.; Puchovsky, M.; Torero, J.L.; Watts, J.M.; Wieczorek, C.J. *SFPE Handbook of Fire Protection Engineering*, 5th ed.; Hurley, M.J., Ed.; Springer: New York, NY, USA; Berlin/Heidelberg, Germany; Dordrecht, The Netherlands; London, UK, 2019.

MDPI
St. Alban-Anlage 66
4052 Basel
Switzerland
Tel. +41 61 683 77 34
Fax +41 61 302 89 18
www.mdpi.com

Fire Editorial Office
E-mail: fire@mdpi.com
www.mdpi.com/journal/fire

www.ingramcontent.com/pod-product-compliance
Lightning Source LLC
Chambersburg PA
CBHW061521210926
51850CB00008B/1998
9783036573175